Lecture Notes in Statistics

Volume 222

Series Editors

Peter Diggle, Department of Mathematics, Lancaster University, Lancaster, UK
Ursula Gather, Dortmund, Germany
Scott Zeger, Baltimore, MD, USA

Lecture Notes in Statistics (LNS) includes research work on topics that are more specialized than volumes in Springer Series in Statistics (SSS). The series editors are currently Peter Diggle, Ursula Gather, and Scott Zeger. Ingram Olkin was an editor of the series for many years.

More information about this series at http://www.springer.com/series/694

Claudia Czado

Analyzing Dependent Data with Vine Copulas

A Practical Guide With R

 Springer

Claudia Czado
Department of Mathematics
Technical University of Munich
Garching, Germany

ISSN 0930-0325   ISSN 2197-7186   (electronic)
Lecture Notes in Statistics
ISBN 978-3-030-13784-7   ISBN 978-3-030-13785-4   (eBook)
https://doi.org/10.1007/978-3-030-13785-4

Library of Congress Control Number: 2019935152

This Springer imprint is published by the registered company Springer Nature Switzerland AG.
The registered company address is: Gewerbestrasse 11, 6330 Cham, Switzerland

Preface

This book is written for graduate students and researchers, who are interested in using copula-based models for multivariate data structures. It provides a step-by-step introduction to the class of vine copulas and their statistical inference. This class of flexible copula models has become very popular in the past years for many applications in diverse fields such as finance, insurance, hydrology, marketing, engineering, chemistry, aviation, climatology, and health.

The popularity of vines copulas is due to the fact, that it allows in addition to the separation of margins and dependence by the copula approach, tail asymmetries, and separate multivariate component modeling. This is accommodated by constructing multivariate copulas using only bivariate building blocks, which can be selected independently. These building blocks are glued together to valid multivariate copulas by appropriate conditioning. Thus, also the term pair copula construction was coined by Aas et al. (2009). This approach allows for flexible and tractable dependence models in dimensions of several hundred, thus providing a long-desired extension of the elliptical and Archimedean copula classes. It forms the basis of new approaches in risk, reliability, spatial analysis, simulation, survival analysis, and data mining to name a few.

Books for the experts already exist. The book by Kurowicka and Cooke (2006) was the first book with vines in the context of belief nets, while Kurowicka and Joe (2011) edited a book entirely devoted to vines. The recent book by Joe (2014) provides a current summary focused mostly on theory and gives many references. However there is no book, which provides the basis for applications of the vine approach. This book will fill this gap. It is geared toward students and researcher novel to this area and is focused on statistical estimation and selection methods of vine copulas for data applications. It provides in addition to the necessary background in multivariate statistics and copula theory exploratory data tools and illustrates the concepts using real data examples. Computations are facilitated using the freely available package `VineCopula` of Schepsmeier et al. (2018) package in R (see R Core Team 2017). It also includes numerous exercises and thus can be used as a book for a course in the statistical analysis of vine copulas.

The book starts with a background chapter on multivariate and conditional distributions and copulas. Basic bivariate dependence measures are introduced in Chap. 2. Chapter 3 is devoted to the bivariate building blocks of vine copulas. It includes the parametric classes of elliptical, Archimedean, and extreme value copulas. Their parameter estimation and graphical tools for the identification of sensible bivariate copula models to data are developed. Bivariate conditional copulas are also introduced since the vine copula approach requires the specification of such copulas. In Chap. 4 the decomposition and construction principle of vines is first given in three dimensions and then extended to the special cases of drawable (D-) and canonical (C-) vines. The general case of regular (R-) vines is developed and discussed in Chap. 5. A vine copula has a building plan, which is specified by the vine tree sequence and requires a bivariate copula family and their parameters for each building block. Simulation algorithms are constructed in Chap. 6. Parameter estimation of a specified vine copula including the sequential estimation approach is the topic of Chap. 7. Chapter 8 concentrates on the problem of selecting the bivariate copula family and the selection of the vine tree structure. Classical comparison methods such as Akaike and Schwarz information criteria and the likelihood ratio test proposed by Vuong (1989) are adapted and illustrated in Chap. 9. A case study characterizing the dependence among German assets contained in the DAX index is given in Chap. 10. The final chapter is devoted to recent extensions of vines including advances in estimation, model selection, for special data structures and reviews major applications in finance, life and earth sciences, insurance, and engineering. It closes with an overview of the available software to select, estimate, and visualize vine-based models. The R code of all figures and tables can be found at http://www.statistics.ma.tum.de/personen/claudia-czado/r-code-to-analyzing-dependent-data-with-vinecopulas.

This book would not have been written without the help of many colleagues and students. First, I would like to thank A. Frigessi and K. Aas who exposed me to the wonderful wide world of vines. A special thank you also to H. Joe, R. Cooke, and D. Kurowicka, who helped me in countless conversions to build a deep understanding of this flexible copula class. Most important are, however, the willingness to accompany me and to contribute to this research journey my former and current Ph.D. students E. C. Brechmann, J. Stöber, U. Schepsmeier, D. Schmidl, A. Bauer, L. Gruber, T. Erhardt, D. Kraus, M. Killiches, D. Müller, L. Höhndorf, T. Nagler, N. Barthel, A. Kreuzer and my postdoctoral students A. Min, A. Panagiotelis and C. Almeida. Vine-based data modeling would have never been so successful without the availability of software. In this area, I have to thank J. Dißmann, U. Schepsmeier, J. Stöber, E. C. Brechmann, B. Gräler, T. Nagler, T. Erhardt, T. Vatter and H. Joe for their computational contributions to the package VineCopula.

Garching, Germany   Claudia Czado

Contents

List of Figures

List of Tables

List of Examples

List of Exercises

Multivariate Distributions and Copulas

1.1 Univariate Distributions

Before we describe multivariate distribution, we review some notation and characteristics of *univariate* distributions. In general, we use capital letters for random variables and small letters for observed values, i.e., we write $X = x$. Here, we only consider absolutely continuous or discrete distributions, and therefore corresponding (conditional) densities or probability mass functions exist. In both cases, we use the letter f for densities or probability mass functions and the letter F for the corresponding distribution function.

In the first course on statistics, standard parametric distributions for univariate continuous variables are introduced (see, for example, Johnson et al. 1995). These include the uniform, exponential, and gamma distribution in addition to the normal and Student's t distribution. The book by Klugman et al. (2012) contains more flexible distribution classes for positive random variables used for loss data. Skewed versions of the normal and Student's t distribution are introduced and discussed in Azzalini and Capitanio (2014). Due to their importance and use in statistics, we only introduce the univariate normal and Student's t distribution and leave the reader to consult the above references for other univariate distributions.

Ex 1.1 (*Univariate Normal and Student's t-distribution*) The density of a *univariate normal* distribution with mean $\mu \in \mathbb{R}$ and variance $\sigma^2 > 0$ is given by

$$f(x; \mu, \sigma^2) = \frac{1}{\sqrt{2\pi\sigma^2}} \exp\left\{ -\frac{1}{\sigma^2}(x - \mu)^2 \right\}. \qquad (1.1)$$

We denote a random variable X with a normal distribution with mean μ and variance σ^2 by $X \sim N(\mu, \sigma^2)$. For the standard normal distribution $N(0, 1)$, the

---

© Springer Nature Switzerland AG 2019
C. Czado, *Analyzing Dependent Data with Vine Copulas*, Lecture Notes in Statistics 222, https://doi.org/10.1007/978-3-030-13785-4_1

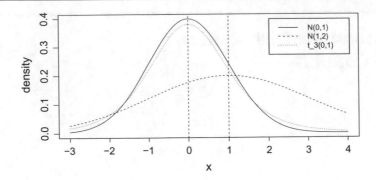

Fig. 1.1 Univariate densities: normal and Student's t distribution with $\nu = 3$

density and distribution function are abbreviated as $\phi(\cdot)$ and $\Phi(\cdot)$, respectively. In the case of the univariate Student's t distribution with mean $\mu \in \mathbb{R}$, scale parameter $\sigma^2 >$ and degree of freedom parameter $\nu > 0$ the density is given by

$$f_\nu(x; \mu, \sigma^2) := \frac{\Gamma\left(\frac{\nu+1}{2}\right)}{\Gamma\left(\frac{\nu}{2}\right)\sqrt{(\pi\nu)}\sigma} \left\{1 + \left(\frac{x-\mu}{\sigma}\right)^2 \frac{1}{\nu}\right\}^{-\frac{\nu+1}{2}}. \qquad (1.2)$$

We denote a random variable X with *Student's t distribution* by $X \sim t_\nu(\mu, \sigma^2)$. For $\nu > 2$, the mean and variance are given by

$$E(X) = \mu \text{ and } \mathrm{Var}(X) = \frac{\nu}{\nu - 2}\sigma^2.$$

The density of a Student's t distribution with mean $\mu = 0$ and scale parameter $\sigma^2 = 1$ is abbreviated by t_ν and is called univariate standard t distribution. The corresponding distribution function is denoted by T_ν. In Fig. 1.1, we illustrate the well-known shape of these univariate distributions. We see that the Student's t distribution has heavier tails than the corresponding normal distribution. Since in financial data often heavy tails are observed, the Student's t distribution is often used as error distribution.

In general, the parameters of the distribution of a random variable X are unknown and have to be estimated. Often estimation is based on a sample $x_1, ..., x_n$ of independent identically distributed (i.i.d) observations from X. If we use a parametric model for X with parameter vector $\boldsymbol{\theta}$, i.e., we assume $X \sim f(\cdot; \boldsymbol{\theta})$, the parameter vector $\boldsymbol{\theta} \in \Theta$ is commonly estimated by *maximum likelihood*, i.e.,

$$\hat{\boldsymbol{\theta}} := \arg\max_{\boldsymbol{\theta} \in \Theta} \prod_{i=1}^{n} f(x_i; \boldsymbol{\theta}). \tag{1.3}$$

Here, Θ denotes the corresponding parameter space for $\boldsymbol{\theta}$ and the function to be maximized in (1.3) is called the *likelihood function*.

The associated distribution function $F(\cdot; \boldsymbol{\theta})$ is then estimated by $F(\cdot; \hat{\boldsymbol{\theta}})$. If one does not want to make the assumption of a parametric statistical model, the *univariate empirical distribution* is often utilized.

> **Definition 1.1** (*Empirical distribution function*) Let $x_1, \ldots, x_n$ be an i.i.d. sample from a distribution function F, then the empirical distribution function is defined as
>
> $$\hat{F}(x) := \frac{1}{n+1} \sum_{i=1}^{n} 1_{\{x_i \leq x\}} \text{ for all } x.$$
>
> Division by $n+1$ instead of n is used to avoid boundary problems of the estimator $\hat{F}(x)$.

Remark 1.2 (*Ranks and empirical distributions*) Let R_i be the *rank* of observation x_i, i.e., $R_i = k$ if the observation x_i is the kth largest observation among the observations $x_1, \ldots, x_n$. In this case, it follows that $\hat{F}(x_i) = \frac{R_i}{n+1}$ for $i = 1, \ldots, n$.

To characterize the dependence between several random variables, we need to standardize random variables. For this, we use the *probability integral transform*.

> **Definition 1.3** (*Probability integral transform*) If $X \sim F$ is a continuous random variable and x is an observed value of X, then the transformation $u := F(x)$ is called the probability integral transform (PIT) at x.

Remark 1.4 (*Distribution of the probability integral transform*) If $X \sim F$ then $U := F(X)$ is uniformly distributed, since

$$P(U \leq u) = P(F(X) \leq u) = P(X \leq F^{-1}(u)) = F(F^{-1}(u)) = u$$

holds for every $u \in [0, 1]$. If F is estimated by $F(\cdot; \hat{\theta})$ or by the empirical distribution $\hat{F}$, then this holds only approximately.

Ex 1.2 (Parametric and nonparametric PIT histograms of normal samples)
To illustrate the probability integral transformation, we generate a sample of sizes 100 and 500 of a standard normal random variable and consider the associated PIT samples in Fig. 1.2. We see from the last row of Fig. 1.2 that for a small sample size the parametric PIT histogram can be quite different than the one expected from of a uniform distribution (uniformly flat). This changes if a larger sample size is used. A sample size of 100 is enough when the empirical distribution function is used as seen from the middle row of Fig. 1.2.

1.2 Multivariate Distributions

Multivariate distributions describe the random behavior of several random variables. In this case, we can distinguish between *marginal*, *joint*, and *conditional* distributions arising from the multivariate distribution. Generally, we denote random vectors in d dimensions in bold letters and subsets of random vectors by X_D for the subset D of $\{1, ..., d\}$. Marginal distribution functions and densities have a subscript j, while conditional distributions of X_j given X_k are indicated by subscripts $j|k$.

Definition 1.5 (*Marginal, joint, and conditional distributions*) For d variate random vector $X = (X_1, ..., X_d)^{\top}$, we use the following notation:

	density function	distribution function
marginal	$f_j(x_j), \ j = 1, ..., d$	$F_j(x_j), \ j = 1, ..., d$
joint	$f(x_1, ..., x_d)$	$F(x_1, ..., x_d)$
conditional	$f_{j|k}(x_j|x_k), \ j \neq k$	$F_{j|k}(x_j|x_k), \ j \neq k$

The class of *elliptical distributions* is a well-known class of multivariate distributions. For a general introduction and extensions see, for example, Genton (2004).

Definition 1.6 (*Elliptical distribution*) The d-dimensional random vector $\mathbf{X}$ has an elliptical distribution if and only if the density function $f(x)$ possesses the representation

$$f(\boldsymbol{x}; \boldsymbol{\mu}, \Sigma) = k_d |\Sigma|^{-\frac{1}{2}} g((\boldsymbol{x} - \boldsymbol{\mu})^{\top} \Sigma^{-1} (\boldsymbol{x} - \boldsymbol{\mu})),$$

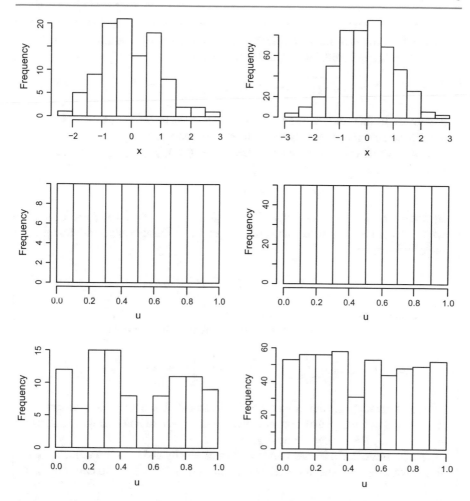

Fig. 1.2 Histograms: (top row) of a standard normal random sample of size 100 (left) and 500 (right) and their associated PIT values using the empirical distribution function (middle row) and the standard normal distribution function (bottom row)

for some constant $k_d \in \mathbb{R}$ only dependent on the dimension d, a mean vector $\boldsymbol{\mu} \in \mathbb{R}^d$, a $\Sigma \in \mathbb{R}^{d \times d}$ symmetric positive definite matrix, and some function $g : \mathbb{R}_0^+ \to \mathbb{R}_0^+$, which is independent of the dimension d.

The *multivariate normal* and the *multivariate Student's t-distribution* are members of this class.

Ex 1.3 (*Multivariate normal distribution*) The multivariate normal distribution (see, for example, Anderson et al. 1958) arises if

$$g(t) := \exp\{-t/2\} \text{ and } k_d = (2\pi)^{-d/2}$$

is chosen. In this case, we say that the random vector $X := (X_1, ..., X_d)^\top$ is multivariate normally distributed with mean vector $\mu := (\mu_1, ..., \mu_d)^\top \in \mathbb{R}^d$ and positive definite covariance matrix $\Sigma = (\sigma_{ij})_{i,j=1,...,d} \in \mathbb{R}^{d \times d}$ and we write

$$X \sim N_d(\mu, \Sigma).$$

In particular, $E(X_i) = \mu_i$ and $\text{Cov}(X_i, X_j) = \sigma_{ij}$ for all $i, j = 1, ..., d$, where σ_{ij} is the (i, j)th element of the matrix Σ. Furthermore, the marginal distributions satisfy

$$X_i \sim N(\mu_i, \sigma_{ii}) \quad \forall i = 1, ..., d,$$

where one often writes $\sigma_i^2 := \sigma_{ii}$ instead.

The density of the multivariate normal distribution is given as

$$f_N(x; \mu, \Sigma) = \frac{1}{(2\pi)^{d/2}} |\Sigma|^{-1/2} \exp\left\{-\frac{1}{2}(x - \mu)^\top \Sigma^{-1}(x - \mu)\right\},$$

where $|A|$ denotes the determinant of the matrix A. The distribution function of a multivariate normal distribution with zero mean vector and correlation matrix R, i.e., $N_d(\mathbf{0}, R)$, we denote by $\Phi_d(\cdots; R)$.

We illustrate the bivariate standard normal density and its contour lines (solid lines) with zero means, unit variances, and a correlation ρ in the two most left panels of Fig. 1.3.

Ex 1.4 (*Conditional distributions of multivariate normal distributions*) Of particular interest are conditional distributions of sub-vectors of a multivariate normally distributed random vector X. Here, we assume a nondegenerate distribution, i.e., the covariance matrix Σ is positive definite. Let $X = (X_1, X_2) \in \mathbb{R}^{d_1+d_2}$, $\mu = (\mu_1^\top, \mu_2^\top)^\top \in \mathbb{R}^{d_1+d_2}$ for $d_1 + d_2 = d$ and partition

$$\Sigma = \begin{pmatrix} \Sigma_{11} & \Sigma_{12} \\ \Sigma_{12}^\top & \Sigma_{22} \end{pmatrix}, \tag{1.4}$$

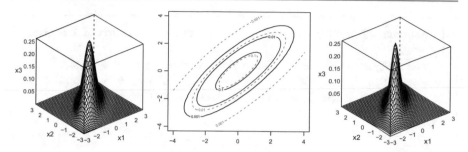

Fig. 1.3 Bivariate densities and contour lines: left: bivariate normal with zero means, unit variances and $\rho = .8$, middle: contour lines for bivariate normal with $\rho = .8$ (solid lines) and bivariate standard Student's t with $\nu = 3, \rho = .8$ (dotted lines) and right: bivariate standard Student's t with $\nu = 8, \rho = .8$

---

where $\Sigma_{11} \in \mathbb{R}^{d_1 \times d_1}$, $\Sigma_{22} \in \mathbb{R}^{d_2 \times d_2}$ and $\Sigma_{12} \in \mathbb{R}^{d_1 \times d_2}$. Then, the conditional distribution of X_2 given $X_1 = x_1$ is

$$X_2 | X_1 = x_1 \sim N_d(\mu_{2|1}, \Sigma_{2|1}), \tag{1.5}$$

where the conditional mean vector is determined by

$$\mu_{2|1} := \mu_2 + \Sigma_{12}^{\top} \Sigma_{11}^{-1} (x_1 - \mu_1),$$

and the conditional covariance matrix by

$$\Sigma_{2|1} := \Sigma_{22} - \Sigma_{12}^{\top} \Sigma_{11}^{-1} \Sigma_{12}.$$

For a proof see Anderson et al. (1958).

Ex 1.5 (Multivariate Student's t distribution) Following Kotz and Nadarajah (2004), the random vector $X = (X_1, ..., X_d)^{\top}$ has *multivariate (or d-variate) Student's t distribution* with $\nu > 0$ degrees of freedom (df), mean vector $\mu \in \mathbb{R}^d$, and scale parameter matrix Σ (a $d \times d$ symmetric and positive definite matrix with elements ρ_{ij}), denoted by $t_d(\nu, \mu, \Sigma)$, if its density is given by

$$f_t(x; \nu; \mu; \Sigma) = \frac{\Gamma(\frac{\nu+d}{2})}{\Gamma(\frac{\nu}{2})(\nu\pi)^{\frac{d}{2}}} |\Sigma|^{-\frac{1}{2}} \left\{ 1 + \frac{1}{\nu}(x - \mu)^{\top} \Sigma^{-1} (x - \mu) \right\}^{-\frac{\nu+d}{2}}. \tag{1.6}$$

The matrix Σ in (1.6) is the correlation matrix of X, i.e., $\mathrm{Cor}(X_i, X_j) = \rho_{ij}$. The Student's t distribution is said to be central if $\mu = \mathbf{0}$; otherwise, it is said to be noncentral. Note that for $\nu > 2$ the variance covariance matrix of X is given by $\frac{\nu}{\nu-2}\Sigma$.

The multivariate Student's t distribution also allows the following *stochastic representation*: If $X = (X_1, ..., X_d)^\top \sim t_d(\nu, \mu, \Sigma)$ with $\nu > 2$ and if $Y = (Y_1, ..., Y_d)^t \sim N_d(\mathbf{0}, \text{Cov})$ with covariance matrix $\text{Cov} = (\text{cov}_{ij})_{i,j=1...d}$, where

$$\text{cov}_{ij} := \begin{cases} \sigma^2 & \text{for } i = j \\ \sigma^2 \rho_{ij} & \text{otherwise} \end{cases}$$

and $\nu S^2/\sigma^2 \sim \chi_\nu^2$ with df ν, independent of Y, then the following identity in distribution

$$X \overset{D}{=} S^{-1}Y + \mu \tag{1.7}$$

holds. Here, χ_ν^2 denotes a Chi-square distribution with ν degrees of freedom. The representation (1.7) implies that the conditional distribution of X given $S^2 = s^2$ is given by

$$X|S^2 = s^2 \sim N_d\left(\mu, \frac{1}{s^2}\text{Cov}\right). \tag{1.8}$$

Finally choosing

$$g(t) := \left(1 + \frac{t}{\nu}\right)^{-(\nu+d)/2} \quad \text{and } k_d := \frac{\Gamma(\frac{\nu+d}{2})}{\Gamma(\frac{\nu}{2})}$$

and using (1.6) shows that the multivariate Student's t distribution is a member of the class of elliptical distributions.

Ex 1.6 (Bivariate standard Student's t distribution) The *bivariate standard Student's t distribution* is a special case of (1.6) with $d = 2$, zero mean vector μ, and scale parameter matrix

$$\Sigma_\rho = \begin{bmatrix} 1 & \rho \\ \rho & 1 \end{bmatrix}. \tag{1.9}$$

The associated density is given by

$$t(x_1, x_2; \nu, \rho) = \frac{\Gamma(\frac{\nu+2}{2})(1 - \rho^2)^{-1/2}}{\Gamma(\frac{\nu}{2})(\nu\pi)} \left\{1 + \frac{1}{\nu}\frac{x_1^2 - 2x_1x_2\rho + x_2^2}{1 - \rho^2}\right\}^{-\frac{\nu+2}{2}}. \tag{1.10}$$

The density and the contour lines for a bivariate Student's t distribution with zero means, scale matrix with correlation $\rho = .8$ and $df = 3$ are given in the left panel and as dotted lines in middle panel of Fig. 1.3. We see that for small contour levels, the Student's t-density is larger than the Gaussian density with the zero means, unit variances, and correlation ρ, thus the bivariate standard Student's t distribution has heavier tails than the bivariate Gaussian distribution with zero means, unit variances, and correlation ρ. Recall that the marginal variances of these distributions do not match.

Ex 1.7 (Conditional distributions of multivariate t) As, for example, shown in Kotz and Nadarajah (2004) if $X = (X_1, ..., X_d)^\top \sim t_d(\nu, \mathbf{0}, \Sigma)$ and the partition of Σ for $X = (X_1, X_2)^\top$ as in (1.4) holds, then the conditional density of X_2 given X_1 is given by

$$f_{X_2|X_1}(x_2|x_1) = \frac{\Gamma(\frac{\nu+d}{2})}{(\pi\nu)^{\frac{d_1}{2}}\Gamma(\frac{\nu+d_1}{2})} \cdot \frac{|\Sigma_{11}|^{\frac{1}{2}}}{|\Sigma|^{\frac{1}{2}}} \cdot \frac{[1 + (1/\nu)x_1^\top \Sigma_{11}^{-1} x_1]^{\frac{\nu+d_1}{2}}}{[1 + (1/\nu)x^\top \Sigma^{-1} x]^{\frac{\nu+d}{2}}}.$$
(1.11)

More generally, if $X = (X_1, ..., X_d)^\top \sim t_d(\nu, \boldsymbol{\mu}, \Sigma)$ for Σ positive definite then the conditional distribution of X_2 given $X_1 = x_1$ is again multivariate $t_{d_2}(\nu_{2|1}, \boldsymbol{\mu}_{2|1}, \Sigma_{2|1})$ with

$$\boldsymbol{\mu}_{2|1} := \boldsymbol{\mu}_2 + \Sigma_{12}^\top \Sigma_{11}^{-1} (x_1 - \boldsymbol{\mu}_1)$$

$$\Sigma_{2|1} := \frac{\nu + (x_1 - \boldsymbol{\mu}_1)^\top \Sigma_{11}^{-1} (x_1 - \boldsymbol{\mu}_1)}{\nu + d_1}(\Sigma_{22} - \Sigma_{12}^\top \Sigma_{11}^{-1} \Sigma_{12}) \quad (1.12)$$

$$\nu_{2|1} := \nu + d_1.$$

More details can be found in Kotz and Nadarajah (2004).

Ex 1.8 (Conditional distribution of bivariate standard Student's t-distribution) For the bivariate standard Student's t distribution introduced in Example 1.6, we use Eqs. (1.12) to determine the conditional parameters of X_1 given $X_2 = x_2$ as

$$\mu_{1|2} := \rho x_2, \sigma_{1|2}^2 := \Sigma_{1|2} = \frac{(\nu + x_2^2)(1 - \rho^2)}{\nu + 1} \text{ and } \nu_{1|2} := \nu + 1. \quad (1.13)$$

In particular, the conditional density $t_{1|2}(x_1|x_2; \nu, \rho)$ is then a univariate Student's t-density with the parameters given by (1.12) and can be expressed as

$$t_{1|2}(x_1|x_2;\nu,\rho) = \frac{\Gamma(\frac{\nu+2}{2})}{\Gamma(\frac{\nu+1}{2})[\pi(\nu+x_2)(1-\rho^2)]^{\frac{1}{2}}}\left[1+\frac{(x_1-\rho x_2)^2}{(1-\rho^2)(\nu+x_2^2)}\right]^{-\frac{\nu+1}{2}}.$$

$$(1.14)$$

Now we extend the notion of an univariate empirical distribution to the multivariate case, which is used if parametric assumptions are to be avoided.

Definition 1.7 (*Multivariate empirical distribution*) Suppose $x_i = (x_{1i}, ..., x_{di})$ is an i.i.d. sample of size n from the d-dimensional distribution F, then the *multivariate empirical distribution function* is defined as

$$\hat{F}(x_1, ...x_d) := \frac{1}{n+1}\sum_{i=1}^{n}1_{\{x_{1i}\leq x_1,...,x_{di}\leq x_d\}} \text{ for all } x := (x_1, ..., x_d)^\top \in \mathbb{R}^d.$$

1.3 Features of Multivariate Data

To explore the shape of univariate data, classical visualization tools such as histograms are sufficient. For multivariate data, these marginal displays are not sufficient and first exploration of such data includes plots of pairs of variables. However, in pairwise plots, the marginal behavior of the variables is mixed with the dependence structure between the variables. To extract marginal effects on the dependence structure, it is useful to normalize each variable such that it has a standard normal distribution and then to consider associated pairs plots. We illustrate this way of proceeding in a three-dimensional data set.

Ex 1.9 (*Chemical components of wines*) Cortez et al. (2009) analyzed 1599 red wine samples according to several chemical components including

- acf = fixed acidity
- acv = volatile acidity
- acc = citric acid

We now investigate their marginal distributions using histograms. The histograms for each of the three components are given in top row of Fig. 1.4.

We see that the marginal distributions are non-normal and different for each component. The pairs plots in the middle row of Fig. 1.4 show the joint effects of the margins and the dependence between the two variables. In the bottom row of Fig. 1.4, the marginal effects are removed by standardizing each variable to have standard normal margins. This is done by first using the probability integral transform based on the empirical distribution and then applying the quantile function of the standard normal distribution. The resulting pairs plots of the transformed variables in the bottom row show that there is dependence between the variables since the data clouds are not balls. In particular, the joint occurrence of small values for two normalized variables acf and acc is different than the occurrence of large values. This indicates nonsymmetric dependence structures in this pair of variables.

The findings of Example 1.9 are common for multivariate data, i.e., different marginal distributions and nonsymmetric dependencies between some pairs of variables.

Standard parametric distributions such as the multivariate normal or Student's t distribution cannot accommodate different types of marginal distribution, since the marginal distributions belong to the same distributional class (see, for example, Anderson et al. 1958). Further, they are symmetric distributions.

The observed nonsymmetric dependence between pairs of variables can also be an indication of heavy tail dependencies between these pairs. Under tail dependence, we understand the dependence, when two variables take on extremely large or small values. In particular, the multivariate normal distribution is not heavy tail-dependent, while the multivariate Student's t distribution has only a single parameter to govern tail dependence. This is more discussed in Sect. 2.3.

In the following section, we introduce an approach which is more suitable for modeling multivariate data.

1.4 The Concept of a Copula and Sklar's Theorem

The copula approach to multivariate data allows to model the margins individually. We have seen the necessity of this already in Example 1.9. It is also clear that the shape of scatter plots between two variables depends on the scale of the variables. Thus, we like to standardize each variable to see if there is dependence between the standardized variables. As standardization tool, the probability integral transform of Definition 1.3 can be utilized. So we want to characterize the dependence between random variables with the common marginal distribution given by the uniform distribution. This approach separates the dependence between the components from the marginal distributions. For this, the dependence between random variables with

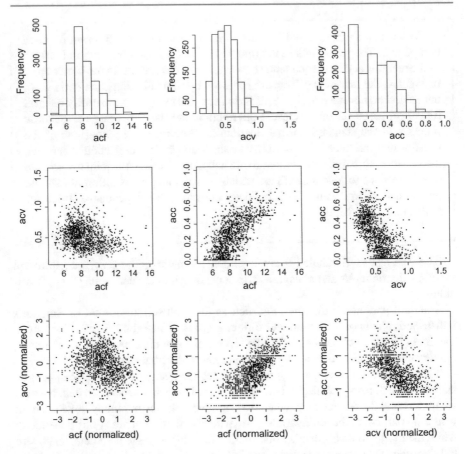

Fig. 1.4 WINE3 : top row: marginal histograms left: acf, middle: acv, right: acc; middle row: pairs plots left: acf versus acv, middle: acf versus acc, right: acv versus acc. bottom row: normalized pairs plots left: acf versus acv, middle: acf versus acc, right: acv versus acc

uniform margins has to be modeled by a corresponding joint distribution function which is called a *copula*.

Definition 1.8 (*Copula and copula density*)

- A d-dimensional copula C is a multivariate distribution function on the d-dimensional hypercube $[0, 1]^d$ with uniformly distributed marginals.
- The corresponding copula density for an absolutely continuous copula we denote by c can be obtained by partial differentiation, i.e., $c(u_1, ..., u_d) := \frac{\partial^d}{\partial u_1 ... \partial u_d} C(u_1, ..., u_d)$ for all $\boldsymbol{u}$ in $[0, 1]^d$.

Sklar (1959) proved the following fundamental representation theorem for multivariate distributions in terms of their marginal distributions and a corresponding copula.

Theorem 1.9 (Sklar's Theorem) *Let X be a d-dimensional random vector with joint distribution function F and marginal distribution functions F_i, $i = 1, \ldots, d$, then the joint distribution function can be expressed as*

$$F(x_1, ..., x_d) = C(F_1(x_1), ..., F_d(x_d)) \tag{1.15}$$

with associated density or probability mass function

$$f(x_1, ..., x_d) = c(F_1(x_1), ..., F_d(x_d)) f_1(x_1)...f_d(x_d) \tag{1.16}$$

for some d-dimensional copula C with copula density c. For absolutely continuous distributions, the copula C is unique.

The inverse also holds: the copula corresponding to a multivariate distribution function F with marginal distribution functions F_i for $i = 1, \ldots, d$ can be expressed as

$$C(u_1, ..., u_d) = F(F_1^{-1}(u_1), ..., F_d^{-1}(u_d)) \tag{1.17}$$

and its copula density or probability mass function is determined by

$$c(u_1, ..., u_d) = \frac{f(F_1^{-1}(u_1), ..., F_d^{-1}(u_d))}{f_1(F_1^{-1}(u_1)) \cdots f_d(F_d^{-1}(u_d))}. \tag{1.18}$$

Proof For absolutely continuous distribution, we see that $C(u_1, ..., u_d)$ is the distribution function of $(U_1, ..., U_d)$ for $U_j := F_j(X_j)$, $j = 1, ..., d$. The proof for the general case can be found in the book by Nelsen (2006). $\square$

Remark 1.10 (Usage of Sklar's Theorem) There are two major applications of Sklar's Theorem.

- **Estimation of the dependence between the standardized variables**: The dependence between standardized variables can be characterized by the copula. In practical data applications, we need to have an i.i.d sample $x_i = (x_{i1}, \ldots, x_{id})^\top$ for $i = 1 \ldots, n$ available. For this, we construct *pseudo-copula data* using an estimated probability integral transform by setting

$$u_{ij} := F_j(x_{ij}; \hat{\boldsymbol{\theta}}_j) \text{ for } j = 1, \ldots, d \text{ and } i = 1, \ldots, n,$$

where $\hat{\theta}_j$ is a parameter estimate of θ_j in a parametric statistical model for the jth marginal. The empirical distribution function of Definition 1.1 can alternatively be used. Now the dependence between the standardized variables u_{ij} can be utilized to find an appropriate model for the copula.

• **Construction of multivariate distributions**: Sklar's Theorem also allows to combine arbitrary marginal distributions together with a copula or copula density to build new multivariate distribution and density functions by using (1.15) and (1.16), respectively.

Ex 1.10 (Chemical components of red wines on the copula scale) We again consider the data of Example 1.9. First, we transform the data using the empirical distribution function $\hat{F}$ for each variable separately to get approximately uniformly distributed data for each variable. The resulting data is the pseudo-copula data as discussed in Remark 1.10. The corresponding histograms and scatter plots for each pair of variables of the copula data are given in Fig. 1.5, showing that the marginal distributions are now approximately uniform and the bivariate dependence structure of the pseudo-data now characterizes the dependence on the copula scale. This is in contrast to the scatter plots on the original scale (compare to Fig. 1.4), where the effects of the marginal distribution and the dependence are not separated. For further discussion, see also Sect. 1.8. Additionally, we see from the lower middle panel of Fig. 1.5 that high values of acf and acc more often occur together than low values. This further indicates the presence of asymmetric tail dependence as discussed in Example 1.9.

For theoretical results on copulas, the following general *bounds* are useful.

Theorem 1.11 (Fréchet–Hoeffding bounds) *Let C be a d-dimensional copula. Then for every* $\boldsymbol{u} \in [0, 1]^d$,

$$W^d(\boldsymbol{u}) \leq C(\boldsymbol{u}) \leq M^d(\boldsymbol{u}), \tag{1.19}$$

where $W^d(\boldsymbol{u}) := \max(u_1 + \ldots + u_d - d + 1, 0)$ *and* $M^d(\boldsymbol{u}) := \min(u_1, \ldots, u_d)$.

It can be shown that the upper bound M^d is a copula, while the lower bound W^d is a copula only for $d = 2$. For a proof of Theorem 1.11, see the books by Joe (1997) or Nelsen (2006).

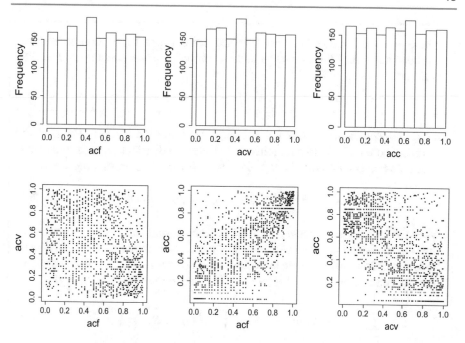

Fig. 1.5 WINE3: top row: marginal histograms of the copula data (left: acf, middle: acv, right: acc), bottom row: pairs plots of the copula data (left: acf versus acv, middle: acf versus acc, right: acv versus acc)

1.5   Elliptical Copulas

We now discuss the copulas derived from elliptical distributions such as the multivariate normal and Student's t distributions.

Ex 1.11 (Bivariate Gaussian copula) The *bivariate Gaussian copula* can be constructed using a bivariate normal distribution with zero mean vector, unit variances, and correlation ρ and applying the inverse statement (1.17) of Sklar's Theorem to obtain

$$C(u_1, u_2; \rho) = \Phi_2(\Phi^{-1}(u_1), \Phi^{-1}(u_2); \rho),$$

where $\Phi(\cdot)$ is the distribution function of a standard normal $N(0, 1)$ distribution and $\Phi_2(\cdot, \cdot; \rho)$ is the bivariate normal distribution function with zero means, unit variances, and correlation ρ. The corresponding copula density can be expressed as

$$c(u_1, u_2; \rho) = \frac{1}{\phi(x_1)\phi(x_2)} \frac{1}{\sqrt{1-\rho^2}} \exp\left\{-\frac{\rho^2(x_1^2 + x_2^2) - 2\rho x_1 x_2}{2(1-\rho^2)}\right\},$$

where $x_1 := \Phi^{-1}(u_1)$ and $x_2 := \Phi^{-1}(u_2)$. Here, Eq. (1.18) of Sklar's theorem has been utilized. A three-dimensional plot of the Gaussian copula with $\rho = .88$ is in the left panel of Fig. 1.6.

Ex 1.12 (Multivariate Gaussian copula) Applying the inverse statement of Sklar's theorem (1.17) to the multivariate Gaussian distribution with zero mean vector and correlation matrix R yields the *multivariate Gaussian copula* as

$$C(\boldsymbol{u}; R) = \Phi_R\left(\Phi^{-1}(u_1), ..., \Phi^{-1}(u_d)\right), \qquad (1.20)$$

where Φ^{-1} denotes the inverse of the univariate standard normal cumulative distribution function Φ and $\Phi_d(\cdots; R)$ the multivariate standard normal distribution function with zero mean vector, unit variances, and symmetric positive definite correlation matrix $R \in [-1, 1]^{d \times d}$. The copula density is then given by

$$c(\boldsymbol{u}; R) = |R|^{-\frac{1}{2}} \exp\left\{\frac{1}{2}\boldsymbol{x}^\top (I_d - R^{-1})\boldsymbol{x}\right\}, \qquad (1.21)$$

where $\boldsymbol{x} = (x_1, ..., x_d)^\top \in \mathbb{R}^d$ with $x_i := \Phi^{-1}(u_i)$, $i = 1, ..., d$.

Ex 1.13 (Bivariate Student t copula) The *bivariate Student's t copula* can be constructed using the bivariate Student's t distribution with ν degrees of freedom, zero mean, and correlation ρ as discussed in Example 1.6 and is given by integrating over the bivariate Student's t copula density $t(x_1, x_2; \nu, \rho)$ (compared to (1.10)) expressed using (1.18) of Sklar's Theorem as follows:

$$C(u_1, u_2; \nu, \rho) = \int_0^{u_1} \int_0^{u_2} \frac{t(T_\nu^{-1}(v_1), T_\nu^{-1}(v_2); \nu, \rho)}{t_\nu(T_\nu^{-1}(v_1))t_\nu(T_\nu^{-1}(v_2))} dv_1 dv_2$$

$$= \int_{-\infty}^{b_1} \int_{-\infty}^{b_2} t(x_1, x_2; \nu, \rho) \, dx_1 \, dx_2, \qquad (1.22)$$

where the variable transformation $b_1 := T_\nu^{-1}(u_1)$ and $b_2 := T_\nu^{-1}(u_2)$ is used. For reference, we also give the expression of the bivariate t copula density

$$c(u_1, u_2; \nu, \rho) = \frac{t(T_\nu^{-1}(v_1), T_\nu^{-1}(v_2); \nu, \rho)}{t_\nu(T_\nu^{-1}(v_1))t_\nu(T_\nu^{-1}(v_2))}. \qquad (1.23)$$

The bivariate Student's t copula density for $\nu = 3$ and $\nu = 8$ and association parameter $\rho = .88$ is illustrated in the middle and left panel of Fig. 1.6.

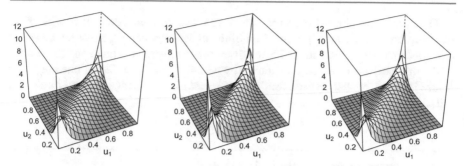

Fig. 1.6 Bivariate copula densities: left: Gauss, middle: Student's t with $\nu = 3$, $\rho = .88$ and right: Student's t with $\nu = 8$, $\rho = .88$

Ex 1.14 (*Multivariate Student's t copula*) Another widely used elliptical copula is the *multivariate t copula* (see also Demarta and McNeil 2005) which is derived from the multivariate Student's t distribution and is given as

$$C(\boldsymbol{u}; R, \nu) = T_{R,\nu}\left(T_\nu^{-1}(u_1), ..., T_\nu^{-1}(u_d)\right), \qquad (1.24)$$

where $T_{R,\nu}$ denotes the distribution function of the multivariate standard Student's t distribution with scale parameter matrix $R \in [-1, 1]^{d \times d}$ and $\nu > 0$ degrees of freedom. The bivariate case was discussed in Example 1.6. Further T_ν^{-1} denotes the inverse of the distribution function T_ν of the univariate standard Student's t distribution with ν degrees of freedom.

1.6 Empirical Copula Approximation

In statistical analyses, we only have multivariate observations available and we need to make inference about the underlying copula. For this, we introduce a data-driven approximation suitable for continuous random vectors of the underlying copula based on a bivariate copula data sample. It utilizes the multivariate empirical distribution introduced in Definition 1.7 for copula data.

Definition 1.12 (*Empirical copula approximation for d = 2*) For the bivariate copula sample $\{u_{1i}, u_{2i}, i = 1, ..., n\}$, the *empirical copula approximation* is defined as

$$\hat{C}(u_1, u_2) := \frac{1}{n+1} \sum_{i=1}^{n} 1_{\{u_{1i} \le u_1, u_{2i} \le u_2\}} \text{ for all } 0 \le u_1, u_2 \le 1. \qquad (1.25)$$

Of course, the general case for $d > 2$ can also be considered. It is a bivariate distribution but only an approximation of the underlying copula, since the marginal distributions associated with (1.25) are discrete and only approximate the continuous uniform distribution on the unit interval when $n < \infty$. In addition, one has to be aware that they cannot be used for the assessment of the dependence structure in the tails.

1.7 Invariance Properties of Copulas

Copulas are *invariant* with respect to strictly increasing transformations of the margins. In particular, the following Lemma holds.

> **Lemma 1.13** (Invariance of copulas) *Let $Y_i := H_i(X_i), i = 1, ..., d$ for H_i strictly increasing functions, then the copula of $Y = (Y_1, ..., Y_d)^\top$ denoted by C_Y agrees with the copula C_X of the underlying random vector $X = (X_1, ..., X_d)^\top$.*

Proof The marginal (inverse) distribution functions F_{Y_i} $(F_{Y_i}^{-1})$ and the joint distribution function F_Y of Y can be determined as

$$F_{Y_i}(y_i) = P(Y_i \leq y_i) = P(X_i \leq H_i^{-1}(y_i)) = F_{X_i}(H_i^{-1}(y_i))$$
$$F_{Y_i}^{-1}(u_i) = H_i(F_{X_i}^{-1}(u_i))$$
$$F_Y(y) = F_X(H_1^{-1}(y_1), ..., H_d^{-1}(y_d)).$$

By Sklar's Theorem using Eq. (1.17), it follows that

$$
\begin{aligned}
C_Y(u) &= F_Y(F_{Y_1}^{-1}(u_1), ..., F_{Y_d}^{-1}(u_d)) \\
&= F_X(H_1(F_{Y_1}^{-1}(u_1)), ..., H_d(F_{Y_d}^{-1}(u_d))) \\
&= F_X(H_1(H_1^{-1}(F_{X_1}^{-1}(u_1))), ..., H_d(H_d^{-1}(F_{X_d}^{-1}(u_d)))) \\
&= F_X(F_{X_1}^{-1}(u_1), ..., F_{X_d}^{-1}(u_d)) \\
&= C_X(u),
\end{aligned}
$$

thus the two copulas C_Y and C_X agree. □

Remark 1.14 (*Invariance for Gaussian and Student's t copulas*) A consequence of Lemma 1.13 is that the copulas corresponding to normal distributions with arbitrary mean vectors and variances coincide and are given as in Example 1.12. The same holds for the copulas corresponding to arbitrary multivariate Student's t distributions.

1.8 Meta Distributions

A distribution which is constructed by an arbitrary copula and arbitrary marginal distributions is called a *meta distribution*.

Ex 1.15 (*Meta-distribution involving a Gaussian copula*) To illustrate a meta-distribution, we combine a bivariate Gaussian copula with a Gaussian margin and an exponential margin. In Fig. 1.7, we present scatter plots for a bivariate sample (u_{i1}, u_{i2}) of size $n = 200$ from a bivariate Gaussian copula with $\rho = .88$ and different margins. Either we use $N(0, 1)$/exponential with rate $\lambda = 5$ margins giving the sample (x_{i1}, x_{i2}) or normal $N(0, 1)/N(0, 1)$ margins giving the sample (z_{i1}, z_{i2}). This shows that scatter plots (left or right panel of Fig. 1.7) of the original data are difficult to interpret with regard to their dependence pattern, since both side panels of Fig. 1.7 have the same dependence as illustrated by the middle panel.

Ex 1.16 (*Meta-distribution involving a non-elliptical copula*) Given the definition of a meta-distribution, we can revisit Examples 1.9 and 1.10. Since the scatter plot on the copula scale of acf and acc in Fig. 1.5 shows nonsymmetric dependence pattern, we need to find copula classes other than the elliptical copula families to capture such behavior. This will be done in Chap. 3.

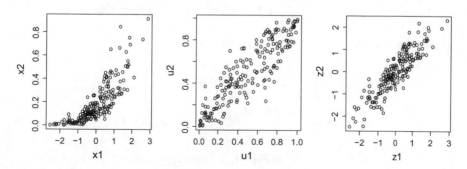

Fig. 1.7 Scatter plots of a bivariate sample ($n = 200$) from Gaussian meta distributions: left: $N(0, 1)$/exponential with rate $\lambda = 5$ margins, middle: uniform[0, 1]/uniform[0, 1] margins, right: $N(0, 1)/N(0, 1)$ margins. The common copula is a bivariate Gauss copula with $\rho = .88$

1.9 Bivariate Conditional Distributions Expressed in Terms of Their Copulas

In the following, we will use the fact that we can express conditional densities in terms of a product of a copula and a marginal density.

Lemma 1.15 (Conditional densities and distribution functions of bivariate distributions in terms of their copula) *The conditional density and distribution function can be rewritten as*

$$f_{1|2}(x_1|x_2) = c_{12}(F_1(x_1), F_2(x_2)) f_2(x_2)$$

$$F_{1|2}(x_1|x_2) = \frac{\partial}{\partial u_2} C_{12}(F_1(x_1), u_2)|_{u_2 = F_2(x_2)}$$

$$=: \frac{\partial}{\partial F_2(x_2)} C_{12}(F_1(x_1), F_2(x_2)).$$

Proof Using the definition a conditional density and Eq. (1.16), we have

$$f_{1|2}(x_1|x_2) = \frac{f_{12}(x_1, x_2)}{f_2(x_2)}$$

$$= \frac{c_{12}(F_1(x_1), F_2(x_2)) f_1(x_2) f_2(x_2)}{f_2(x_2)}$$

$$= c_{12}(F_1(x_1), F_2(x_2)) f_1(x_2)$$

$$= \frac{\partial^2 C_{12}(u_1, u_2)}{\partial u_1 \partial u_2}|_{u_1 = F_1(x_1), u_2 = F_2(x_2)} \frac{\partial u_1}{\partial x_1}$$

$$= \frac{\partial}{\partial u_2} \left(\frac{\partial}{\partial x_1} C_{12}(F_1(x_1), u_2) \right)|_{u_2 = F_2(x_2)}.$$

This implies that

$$F_{1|2}(x_1|x_2) = \int_{-\infty}^{x_1} \frac{\partial}{\partial u_2} \left(\frac{\partial}{\partial z_1} C_{12}(F_1(z_1), u_2) \right)|_{u_2 = F_2(x_2)} dz_1$$

$$= \frac{\partial}{\partial u_2} \left(\int_{-\infty}^{x_1} \frac{\partial}{\partial z_1} C_{12}(F_1(z_1), u_2) dz_1 \right)|_{u_2 = F_2(x_2)}$$

$$= \frac{\partial}{\partial u_2} C_{12}(F_1(x_1), u_2)|_{u_2 = F_2(x_2)}. \qquad \square$$

Lemma 1.15 can also be applied to the bivariate copula distribution C_{12}. In this case, we denote the conditional distribution and density by $C_{1|2}$ and $c_{1|2}$, respectively.

In particular, it follows that

$$C_{1|2}(u_1|u_2) = \frac{\partial}{\partial u_2} C_{12}(u_1, u_2) \ \forall u_1 \in [0, 1]. \tag{1.26}$$

The relationship between $F_{1|2}$ and $C_{1|2}$ using Lemma 1.15 is therefore given by

$$F_{1|2}(x_1|x_2) = \frac{\partial}{\partial u_2} C_{12}(F_1(x_1), u_2)|_{u_2 = F_2(x_2)} = C_{1|2}(F_1(x_1)|F_2(x_2)). \tag{1.27}$$

Applying (1.27) yields the following relationship among the inverse function of the conditional distribution functions:

$$F_{1|2}^{-1}(u_1|x_2) = F_1^{-1}(C_{1|2}^{-1}(u_1|F_2(x_2))). \tag{1.28}$$

The conditional distribution function $C_{1|2}$ associated with a copula was denoted by Aas et al. (2009) also as an *h-function*.

Definition 1.16 (*h-functions of bivariate copulas*) The h functions corresponding to a bivariate copula C_{12} are defined for all $(u_1, u_2) \in [0, 1]^2$ as

$$h_{1|2}(u_1|u_2) := \frac{\partial}{\partial u_2} C_{12}(u_1, u_2)$$

$$h_{2|1}(u_2|u_1) := \frac{\partial}{\partial u_1} C_{12}(u_1, u_2).$$

Ex 1.17 (*Conditional distribution function of a bivariate Student's t copula*) We start with the conditional density $c_{1|2}(u_1|u_2; \nu, \rho)$ of U_1 given $U_2 = u_2$, where (U_1, U_2) are distributed according to the bivariate t copula as introduced in Example 1.13. This is given as

$$c_{1|2}(u_1|u_2; \nu, \rho) = c(u_1, u_2; \nu, \rho), \tag{1.29}$$

where we used the fact that the margins of (U_1, U_2) are uniform. Recall that $c(u_1, u_2; \nu, \rho)$ is defined in (1.23). Then, the conditional distribution function $C_{1|2}(u_1|u_2; \nu, \rho)$ of U_1 given $U_2 = u_2$ can be determined using $b_i := T_\nu^{-1}(u_i)$

for $i = 1, 2$ and the variable transformation $x_1 = T_\nu^{-1}(v_1)$ as

$$
\begin{aligned}
C_{1|2}(u_1|u_2; \nu, \rho) &= \int_0^{u_1} c_{1|2}(v_1|u_2; \nu, \rho)dv_1 = \int_0^{u_1} c(v_1, u_2; \nu, \rho)dv_1 \\
&= \int_0^{u_1} \frac{t(T_\nu^{-1}(v_1), T_\nu^{-1}(u_2); \nu, \rho)}{t_\nu(T_\nu^{-1}(v_1))t_\nu(T_\nu^{-1}(u_2))}dv_1 \\
&= \int_{-\infty}^{b_1} \frac{t(x_1, b_2; \nu, \rho)}{t_\nu(b_2)}dx_1 \\
&= \int_{-\infty}^{b_1} t_{1|2}(x_1|b_2; \nu, \rho)dx_1 \\
&= T_{1|2}(b_1|b_2; \nu, \rho) \\
&= T_{\nu+1}\left(\frac{T_\nu^{-1}(u_1) - \rho T_\nu^{-1}(u_2)}{\sqrt{\frac{(\nu+(T_\nu^{-1}(u_2))^2)(1-\rho^2)}{\nu+1}}} \right).
\end{aligned}
\tag{1.30}
$$

Here, $t_{1|2}(x_1|x_2; \nu, \rho)$ is the conditional density of X_1 given $X_2 = x_2$ of a bivariate standard t distribution for (X_1, X_2) defined in (1.14) with corresponding distribution function $T_{1|2}(x_1|x_2; \nu, \rho)$. As shown in Example 1.8, this is a univariate Student's t distribution with the parameters identified in (1.13), which explains the last equality.

Finally, we summarize our notation and findings with regard to bivariate distributions and copulas in Table 1.1.

1.10 Exercises

Exer 1.1
Conditional distribution for bivariate Student's t distributions: Derive from first principles the conditional distribution of X_1 given $X_2 = x_2$ when (X_1, X_2) are jointly Student's t distributed with mean vector $\boldsymbol{\mu}$, scale matrix Σ, and ν degree of freedom. Determine the corresponding expectation and variance.

Exer 1.2
Conditional density and distribution of the bivariate Gaussian copula: Derive the h-functions corresponding to a bivariate Gaussian copula with correlation parameter ρ and graph the surface of $h_{1|2}(u_1|u_2; \rho)$ for $0 < u_1 < 1, 0 < \rho < 1$ for fixed $u_2 = .1$ and $u_2 = .8$, respectively.

Exer 1.3
Density and conditional distributions of the bivariate Clayton copula: The *bivariate*

Table 1.1 Relationships among bivariate absolutely continuous distributions and copulas

Distr. quant.	Original scale	Copula scale	Relationship
Joint distr.	F_{12}	C_{12}	$F_{12}(x_1, x_2) = C_{12}(F_1(x_1), F_2(x_2))$ $C_{12}(u_1, u_2) = F_{12}(F_1^{-1}(u_1), F_2^{-1}(u_2))$
Marg. distr.	F_1 F_2	$C_1(u_1) = u_1$ $C_2(u_2) = u_2$	
Joint den.	$f_{12}(x_1, x_2) = \frac{\partial^2 F_{12}(x_1, x_2)}{\partial x_1 \partial x_2}$	$c_{12}(u_1, u_2) = \frac{\partial^2 C_{12}(u_1, u_2)}{\partial u_1 \partial u_2}$	$f_{12}(x_1, x_2) = c_{12}(F_1(x_1), F_2(x_2)) f_1(x_1) f_2(x_2)$ $c_{12}(u_1, u_2) = \frac{f_{12}(F_1^{-1}(u_1), F_2^{-1}(u_2))}{f_1(F_1^{-1}(u_1)) f_2(F_2^{-1}(u_2))}$
Marg. den.	f_1 f_2	$c_1(u_1) = 1$ $c_2(u_2) = 1$	
Cond. den.	$f_{1\|2}(x_1\|x_2) = \frac{f_{12}(x_1, x_2)}{f_2(x_2)}$	$c_{1\|2}(u_1\|u_2) = \frac{c_{12}(u_1, u_2)}{}$	$f_{1\|2}(x_1\|x_2) = c_{12}(F_1(x_1), F_2(x_2)) f_1(x_1)$
Cond. distr.	$F_{1\|2}(x_1\|x_2) = \frac{\frac{\partial}{\partial x_2} F_{12}(x_1, x_2)}{f_2(x_2)}$	$C_{1\|2}(u_1\|u_2) = \frac{\partial}{\partial u_2} C_{12}(u_1, u_2)$	$F_{1\|2} = \frac{\partial}{\partial u_2} C_{12}(F_1(x_1), u_2)\|_{u_2 = F_2(x_2)} =$ $C_{1\|2}(F_1(x_1)\|F_2(x_2))$
Cond.inv. distr.	$F_{1\|2}^{-1}(u_1\|x_2)$	$C_{1\|2}^{-1}(u_1\|u_2)$	$F_{1\|2}^{-1}(u_1\|x_2) = F_1^{-1}(C_{1\|2}^{-1}(u_1\|F_2(x_2)))$

Clayton copula is defined as

$$C(u_1, u_2|\theta) = (u_1^{-\theta} + u_2^{-\theta} - 1)^{-\frac{1}{\theta}}$$

for $0 \leq \theta < \infty$.

1. Show that the corresponding copula density is given by

$$c(u_1, u_2; \theta) = (1 + \theta)(u_1^{-\theta} + u_2^{-\theta} - 1)^{-\frac{1+2\theta}{\theta}} (u_1 u_2)^{-(\theta+1)}$$

2. Show that the conditional distribution function of U_1 given $U_2 = u_2$ is given by

$$C_{1|2}(u_1|u_2; \theta) = (u_1^{-\theta} + u_2^{-\theta} - 1)^{-\frac{1+\theta}{\theta}} u_2^{-(\theta+1)}$$

3. Derive the inverse of $C_{1|2}(u_1|u_2; \theta)$.

Exer 1.4

Multivariate Burr distribution: A d variate version of the *Burr distribution* was introduced by Takahashi (1965) with density for $x_i \geq 0$ and $i = 1, ..., d$

$$f(x; b, r, p) = \frac{\Gamma(p+d)}{\Gamma(p)} \frac{\prod_{k=1}^{d} r_k b_k x_k^{b_k-1}}{(1 + \sum_{k=1}^{d} r_k x_k^{b_k})^{(p+d)}}, \tag{1.31}$$

where $p \geq 1$, $b = (b_1, \ldots, b_d)^\top$ and $r = (r_1, \cdots, r_d)^\top$ with $b_k \geq 1$ and $r_k > 0$ for $k = 1, \cdots, d$. We denote this distribution by $Burr(d, b, r, p)$. For $d = 2$, show the following:

1. The marginal densities and distribution functions for $k = 1, 2$ are

$$f_k(x_k) = \frac{r_k b_k x_k^{b_k-1}}{(1 + r_k x_k^{b_k})^{p+1}} \text{ and } F_k(x_k) = 1 - (1 + r_k x_k^{b_k})^p.$$

2. The bivariate distribution function is given by

$$F_{12}(x_1, x_2; \boldsymbol{b}, \boldsymbol{r}, p) = F_1(x_1) + F_2(x_2) + (1 + r_1 x_1^{b_1} + r_2 x_2^{b_2})^{-p} - 1.$$

3. The associated copula to F_{12} is independent of $\boldsymbol{b}$ and $\boldsymbol{r}$ and is given by

$$C(u_1, u_2; p) = u_1 + u_2 + ((1 - u_1)^{-\frac{1}{p}} + (1 - u_2)^{-\frac{1}{p}} - 1)^{-p} - 1.$$

4. Derive the density and distribution function of the conditional distribution of X_1 given $X_2 = x_2$.

Exer 1.5

URAN3: *Three-dimensional uranium data*: Perform the same analysis as for the wine data in Examples 1.9 and 1.10 for variables Co, Ti and Sc of the data uranium contained in the R package copula of (Hofert et al. 2017). This data was also considered in Acar et al. (2012) and Kraus and Czado (2017b).

Exer 1.6

ABALONE3: *Three-dimensional abalone data*: Perform the same analysis as for the wine data in Examples 1.9 and 1.10 for variables shucked, viscera, and shell of the data abalone contained in the R package PivotalR from Pivotal Inc. (2017). Restrict your analysis to the female abalone shells.

Exer 1.7

WINE7: *Seven-dimensional red wine data*: The chemical components of red wine data studied in Cortez et al. (2009) considered in Examples 1.9 and 1.10 have more components available. The full data set is available from the UCI Machine Learning Repository under https://archive.ics.uci.edu/ml/datasets/wine+quality. In particular, we have measurements

- acf = fixed acidity
- acv = volatile acidity
- acc = citric acid
- clor = chlorides
- st = total sulfur dioxide
- den = density
- ph = pH value

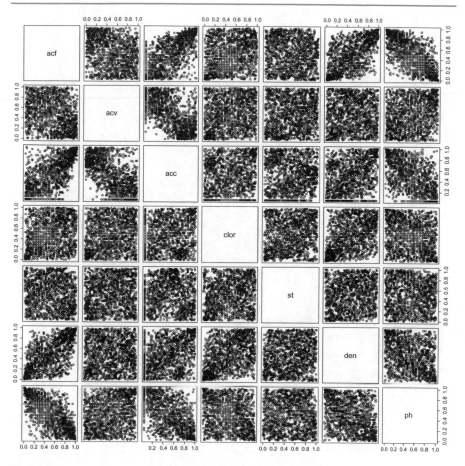

Fig. 1.8 WINE7 : Scatter plots of the seven-dimensional red wine data

available. A plot of all pairwise scatter plots of the pseudo-data obtained using empirical distributions for each margin separately is given in Fig. 1.8.

Based on Fig. 1.8

- Identify which pairs show strong dependence.
- Identify which pairs show an asymmetric tail dependence pattern.
- Identify which pairs show positive and which pairs show negative dependence.

Dependence Measures

There exist several measures for the strength and direction of dependence between two random variables. The most common ones are the Pearson product–moment correlation, Kendall's tau, and Spearman's rho. We give a short introduction to these measures including their estimation based on data. Kendall's tau and Spearman's rho can be expressed in terms of the corresponding copula alone, while this is not always the case for the Pearson product–moment correlation. For joint extreme events, the notion of tail dependence coefficients will be introduced. Finally, the concepts of partial and conditional correlations will be discussed, which will play an important role in the class of multivariate vine copulas.

2.1 Pearson Product–Moment Correlation

The *Pearson product–moment correlation* is a measure of linear dependence with values in the interval $[-1, 1]$. It is not invariant with respect to monotone increasing transformations of the margins. Further, it is not defined for distributions with non-finite second moments (e.g., the Cauchy distribution). Additionally, its value might depend on the marginal distributions of X_1 and X_2 (see Example 3.1 of Kurowicka and Cooke 2006).

Definition 2.1 (*Pearson product–moment correlation*) The Pearson's product–moment correlation coefficient ρ between two random variables X_1 and X_2 with finite second moments is defined as

$$\rho := \rho(X_1, X_2) := \text{Cor}(X_1, X_2) = \frac{\text{Cov}(X_1, X_2)}{\sqrt{\text{Var}(X_1)}\sqrt{\text{Var}(X_2)}}, \quad (2.1)$$

© Springer Nature Switzerland AG 2019
C. Czado, *Analyzing Dependent Data with Vine Copulas*, Lecture Notes
in Statistics 222, https://doi.org/10.1007/978-3-030-13785-4_2

Ex 2.1 (*Zero correlation does not imply independence*) Assume X_1 is a random variable with $E(X_1) = E(X_1^3) = 0$ and consider $X_2 := X_1^2$, then obviously X_1 and X_2 are dependent, however $\text{Cor}(X_1, X_2) = 0$.

From a random sample from (X_1, X_2), the Pearson correlation $\rho(X_1, X_2)$ can be estimated.

Definition 2.2 (*Pearson product–moment correlation estimate*) For a random sample $\{x_{i1}, x_{i2}, i = 1, ..., n\}$ of size n from the joint distribution of (X_1, X_2) the Pearson correlation $\rho(X_1, X_2)$ is estimated by

$$\hat{\rho} := \hat{\rho}(X_1, X_2) := \frac{\sum_{i=1}^{n}(x_{i1} - \bar{x}_1)(x_{i2} - \bar{x}_2)}{\sqrt{\sum_{i=1}^{n}(x_{i1} - \bar{x}_1)^2}\sqrt{\sum_{i=1}^{n}(x_{i2} - \bar{x}_2)^2}}, \tag{2.2}$$

where $\bar{x}_1 := \frac{1}{n}\sum_{i=1}^{n} x_{i1}$ and $\bar{x}_2 := \frac{1}{n}\sum_{i=1}^{n} x_{i2}$ are the corresponding sample means.

Ex 2.2 (*Pearson correlation depends on the margins as well*) We consider again the bivariate sample data from Example 1.15. For the $N(0, 1)/N(0, 1)$ margins the Pearson correlation is estimated as .832, while for uniform margins it is .881 and for the $N(0, 1)/$ exponential with rate $\lambda = 5$ it is .896, respectively. These numbers change to .808, .885, and .893 for a sample size of $n = 10000$. This illustrates that the Pearson correlation ρ is not invariant under monotone transformations.

2.2 Kendall's τ and Spearman's ρ_s

Both *Kendall's tau* and *Spearman's rho* are dependence measures which are rank-based and therefore invariant with respect monotone transformations of the marginals. Their range of values is the interval $[-1, 1]$. Additionally, they can be expressed solely in terms of their associated copula and therefore their value does not depend on the marginal distributions. Often there are closed-form expressions in terms of the copula parameters available.

Kendall's tau, denoted by τ, is defined as the probability of *concordance* minus the probability of *discordance* of two random variables X_1 and X_2.

Definition 2.3 (*Kendall's tau*) The Kendall's τ between the continuous random variables X_1 and X_2 is defined as

$$\tau(X_1, X_2) = P((X_{11} - X_{21})(X_{12} - X_{22}) > 0) - P((X_{11} - X_{21})(X_{12} - X_{22}) < 0), \tag{2.3}$$

where (X_{11}, X_{12}) and (X_{21}, X_{22}) are independent and identically distributed copies of (X_1, X_2).

Nonparametric estimation of Kendall's τ is treated in detail in Chap. 8 of Hollander et al. (2014). In particular to estimate Kendall's τ from a random sample $\{x_{i1}, x_{i2}, i = 1, ..., n\}$ of size n from the joint distribution of (X_1, X_2), we consider all $\binom{n}{2} = \frac{n(n-1)}{2}$ unordered pairs $\boldsymbol{x}_i := (x_{i1}, x_{i2})$ and $\boldsymbol{x}_j := (x_{j1}, x_{j2})$ for $i, j = 1, ..., n$.

Definition 2.4 (*Concordant, discordant, and extra pairs*) The pair $(\boldsymbol{x}_i, \boldsymbol{x}_j)$ is called

- *concordant* if the ordering in $\boldsymbol{x}^1 := (x_{i1}, x_{j1})$ is the same as in $\boldsymbol{x}^2 := (x_{i2}, x_{j2})$, i.e., $x_{i1} < x_{j1}$ and $x_{i2} < x_{j2}$ holds or $x_{i1} > x_{j1}$ and $x_{i2} > x_{j2}$ holds,
- *discordant* if the ordering in $\boldsymbol{x}^1$ is opposite to the ordering of $\boldsymbol{x}^2$, i.e. $x_{i1} < x_{j1}$ and $x_{i2} > x_{j2}$ holds or $x_{i1} > x_{j1}$ and $x_{i2} < x_{j2}$ holds,
- *extra x_1 pair* if $x_{i1} = x_{j1}$ holds,
- *extra x_2 pair* if $x_{i2} = x_{j2}$ holds.

Remark 2.5 (*Characterization of concordance and discordance*) Concordance occurs when $(x_{i1} - x_{j1})(x_{i2} - x_{j2}) > 0$ and discordance occurs when $(x_{i1} - x_{j1})(x_{i2} - x_{j2}) < 0$. In the case of continuous random variables the case $(x_{i1} - x_{j1})(x_{i2} - x_{j2}) = 0$ occurs with probability zero.

Definition 2.6 (*Estimate of Kendall's τ allowing for ties*) Let N_c be the number of concordant pairs, N_d be the number of discordant pairs, N_1 be the number of extra x_1 pairs, and N_2 be the number of extra x_2 pairs of random sample $\{x_{i1}, x_{i2}, i = 1, ..., n\}$ from the joint distribution of (X_1, X_2). Then an *estimate of Kendall's τ* is given by

$$\hat{\tau}_n^* := \frac{N_c - N_d}{\sqrt{N_c + N_d + N_1} \times \sqrt{N_c + N_d + N_2}}. \tag{2.4}$$

The estimate of τ in (2.4) is used in Kurowicka and Cooke (2006), while Genest and Favre (2007) use the simpler estimate assuming no ties.

Definition 2.7 (*Estimate of Kendall's τ allowing for no ties*) An estimate of Kendall's τ for samples of size n without ties is defined as

$$\hat{\tau}_n := \frac{N_c - N_d}{\binom{n}{2}}. \tag{2.5}$$

Both estimates coincide in the case of no ties since the total number of unordered pairs agrees with the sum $N_c + N_d$. Further both estimates do not change if the ranks of the observations are used instead of the data values, thus the estimates are rank based. For large sample sizes n, it is better to use the algorithm of Knight (1966), which is also given as Algorithm 1 on page 269 of Joe (2014).

General results on the asymptotic distribution of $\hat{\tau}_n$ are available. For example, Joe (2014) noted on page 55 that $n^{-1}(\hat{\tau}_n - \tau)$ is asymptotically normal with asymptotic variance given by

$$n\text{Var}(\hat{\tau}_n) \xrightarrow[n \to \infty]{} 16 \int_0^1 \int_0^1 [C(u_1, u_2) + C(1 - u_1, 1 - u_2)]^2 c(u_1, u_2) du_1 du_2, \tag{2.6}$$

where C is the copula associated to (X_1, X_2) with density c. This can be proven using the U-statistic formulation of Hoeffding (1948). From this, the following approximation can be derived.

Theorem 2.8 (Asymptotic approximation of the distribution of Kendall's τ estimate) *The asymptotic distribution of Kendall's τ estimate τ_n based on (2.5) can be approximated for large n by*

$$\sqrt{n}\frac{\hat{\tau}_N - \tau}{4S} \approx N(0, 1), \tag{2.7}$$

where $S^2 := \frac{1}{n}\sum_{i=1}^n (W_i + \tilde{W}_i - 2\bar{W})$ *for* $\bar{W}_i := \frac{1}{n}\#\{j : x_{i1} < x_{j1}, x_{i2} < x_{j2}\}$, $W_i := \frac{1}{n}\#\{j : x_{i1} \le x_{j1}, x_{i2} \le x_{j2}\}$ *and* $\tilde{W} := \frac{1}{n}(W_1 + \cdots + W_n)$.

We now treat the case of independence. In particular if (X_1, X_2) are independent random variables the asymptotic variance specified in (2.6) is $4/9$. Genest and Favre (2007) use direct calculation of the variance of $\hat{\tau}_n$ under the null hypothesis of independence to show

$$\text{Var}(\hat{\tau}_n) = \frac{2(2n + 5)}{9n(n - 1)}.$$

Details on this calculation can be found on page 419 of Hollander et al. (2014). While the mean and variance can be directly determined, the asymptotic normality follows again using the U-statistic formulation of Hoeffding (1948). This allows for the construction of an asymptotic α level test for H_0: X_1 and X_1 are independent versus H_1: not H_0. In particular, reject H_0 versus H_1 if and only if

$$\sqrt{\frac{9n(n-1)}{2(2n+5)}}|\hat{\tau}_n| > z_{1-\alpha/2}, \tag{2.8}$$

where z_β is the β quantile of a standard normal distribution.

We discuss now a second-rank-based dependence measure.

Definition 2.9 (*Spearman's ρ_s or rank correlation*) For continuous random variables X_1 and X_2 with marginal distributions F_1 and F_2, respectively, Spearman's ρ_s or the rank correlation is defined as the Pearson correlation of the random variables $F_1(X_1)$ and $F_2(X_2)$, i.e.,

$$\rho_s := \rho_s(X_1, X_2) = \text{Cor}(F_1(X_1), F_2(X_2)). \tag{2.9}$$

Ex 2.3 (*Relationships between Pearson correlation ρ, Spearman's ρ_s, and Kendall's τ for the bivariate normal distribution*) For the bivariate normal distribution, relationships among all three dependence measures are available. In particular, we have the following results:

$$\rho = 2\sin\left(\frac{\pi}{6}\rho_s\right) \text{ and } \tau = \frac{2}{\pi}\arcsin(\rho).$$

These have been shown in Pearson (1904). We illustrate these relationships in Fig. 2.1. From this we see that Kendall's τ is larger (smaller) than Spearman's ρ_s for $\rho \leq 0$ ($\rho \geq 0$).

We show now that Kendall's τ and Spearman's ρ_s are independent of the marginal specification and thus depend exclusively on the associated copula.

Theorem 2.10 (Kendall's τ and Spearman's ρ_s expressed in terms of the copula) *Let (X_1, X_2) be continuous random variables, then Kendall's τ and Spearman's ρ can be expressed as*

$$\tau = 4\int_{[0,1]^2} C(u_1, u_2)dC(u_1, u_2) - 1 \text{ and } \rho_s = 12\int_{[0,1]^2} u_1u_2dC(u_1, u_2) - 3.$$

$$\tag{2.10}$$

Fig. 2.1 Dependence measures: Kendall's τ (solid) and Spearman's ρ_s (dashed) for a bivariate normal distribution with changing Pearson correlation ρ (dotted line is the $x = y$ axis)

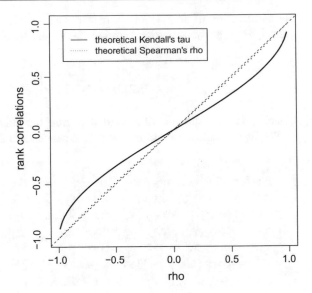

Proof Let (X_{11}, X_{12}) and (X_{21}, X_{22}) be independent copies of (X_1, X_2). For continuous variables, we have

$$P((X_{11} - X_{21})(X_{12} - X_{22}) > 0) = 1 - P((X_{11} - X_{21})(X_{12} - X_{22}) < 0),$$

therefore, $\tau = 2P((X_{11} - X_{21})(X_{12} - X_{22}) > 0) - 1$. Further we have that

$$P((X_{11} - X_{21})(X_{12} - X_{22}) > 0) = P(X_{11} > X_{21}, X_{12} > X_{22}) + P(X_{11} < X_{21}, X_{12} < X_{22}).$$

Now using the transformation $u_1 := F_1(x_1)$ and $u_2 := F_2(x_2)$ gives

$$
\begin{aligned}
P(X_{11} > X_{21}, X_{12} > X_{22}) &= P(X_{21} < X_{11}, X_{22} < X_{12}) \\
&= \int_{-\infty}^{\infty} \int_{-\infty}^{\infty} P(X_{21} < x_1, X_{22} < x_2) dC(F_1(x_1), F_2(x_2)) \\
&= \int_{-\infty}^{\infty} \int_{-\infty}^{\infty} C(F_1(x_1), F_2(x_2)) dC(F_1(x_1), F_2(x_2)) \\
&= \int_0^1 \int_0^1 C(u_1, u_2) dC(u_1, u_2).
\end{aligned}
$$

Similarly, we can show that

$$P(X_{11} < X_{21}, X_{12} < X_{22}) = \int_0^1 \int_0^1 [1 - u_1 - v_1 + C(u_1, u_2)] dC(u_1, u_2).$$

Since C is the distribution function of the random variables $U_1 := F_1(X_1)$ and $U_2 := F_2(X_2)$ and U_i have mean $1/2$, it follows that

$$P(X_{11} < X_{21}, X_{12} < X_{22}) = 1 - \frac{1}{2} - \frac{1}{2} + \int_0^1 \int_0^1 C(u_1, u_2) dC(u_1, u_2)$$

$$= \int_0^1 \int_0^1 C(u_1, u_2) dC(u_1, u_2).$$

This concludes the proof for Kendall's τ. For Spearman's ρ_s, we note that the variance of a uniform distribution on $[0, 1]$ is $1/12$ and thus

$$\rho_s = \mathrm{Cor}(F_1(X_1), F_2(X_2)) = \mathrm{Cor}(U_1, U_2) = \frac{E(U_1 U_2) - \frac{1}{4}}{\frac{1}{12}}$$

$$= 12 E(U_1 U_2) - 3 = 12 \int_0^1 \int_0^1 u_1 u_2 dC(u_1, u_2) - 3.$$

For a random sample $\{x_{i1}, x_{i2}, i = 1, ..., n\}$ of size n from the joint distribution of (X_1, X_2), Spearman's ρ_s is estimated by using the marginal ranks $\{r_{i1}, r_{i2}, i = 1, ..., n\}$, where r_{ij} is the rank of observation x_{ij} for $j = 1, 2$. These are now used in the same way as the original observations for the estimate of the Pearson correlation. This yields the following estimate of *Spearman's* ρ_s

Definition 2.11 (*Estimate of Spearman's* ρ_s) A estimate of Spearman's ρ_s based on a sample (x_{i1}, x_{i2}) of size n with ranks r_{ij} for $j = 1, 2$ is given by

$$\hat{\rho}_s := \hat{\rho}_s(X_1, X_2) := \frac{\sum_{i=1}^n (r_{i1} - \bar{r}_1)(r_{i2} - \bar{r}_2)}{\sqrt{\sum_{i=1}^n (r_{i1} - \bar{r}_1)^2}\sqrt{\sum_{i=1}^n (r_{i2} - \bar{r}_2)^2}}, \qquad (2.11)$$

where $\bar{r}_1 := \frac{1}{n}\sum_{i=1}^n r_{i1}$ and $\bar{r}_2 := \frac{1}{n}\sum_{i=1}^n r_{i2}$ are the corresponding sample rank means.

Ex 2.4 (*Estimated bivariate dependence measures for the wine data of Example* 1.9) The estimated bivariate dependence measures for the wine data are given in Table 2.1. We see that there is positive and negative bivariate dependence in this data set.

Table 2.1 WINE3: Estimated correlation ρ, Kendall's τ, and Spearman's ρ_s for all pairs of variables of the red wine data of Example 1.9

Measure	(acf, acv)	(acf, acc)	(acv, acc)
Pearson ρ	$-.26$	.67	$-.55$
Kendall τ	$-.19$	.48	$-.43$
Spearman ρ_s	$-.28$	.66	$-.61$

2.3 Tail Dependence

For assessing tail dependence an extremal view is taken and the probability of the joint occurrence of extremely small or large values are considered.

Definition 2.12 *(Upper and lower tail dependence coefficient)* The *upper tail dependence coefficient* of a bivariate distribution with copula C is defined as

$$\lambda^{upper} = \lim_{t \to 1^-} P(X_2 > F_2^{-1}(t)|X_1 > F_1^{-1}(t)) = \lim_{t \to 1^-} \frac{1 - 2t + C(t,t)}{1 - t},$$

while the *lower tail dependence coefficient* is

$$\lambda^{lower} = \lim_{t \to 0^+} P(X_2 \leq F_2^{-1}(t)|X_1 \leq F_1^{-1}(t)) = \lim_{t \to 0^+} \frac{C(t,t)}{t}.$$

To illustrate the concept of tail dependence, consider Fig. 2.2 of two variables X_1 and X_2 with standard normal distribution. Here upper tail dependence is driven by the square points, while lower tail dependence is driven by the triangle points. For some parametric copulas, the tail dependence coefficient can be directly computed. We give now two examples.

Ex 2.5 (Tail dependence coefficients of the bivariate Student t copula) For the bivariate Student's t copula C with parameters ρ and ν discussed in Example 1.13 the *upper and lower tail dependence coefficient* are the same and given by

$$\lambda = 2T_{\nu+1}\left(-\sqrt{\nu+1}\sqrt{\frac{1-\rho}{1+\rho}}\right), \tag{2.12}$$

where $T_{\nu+1}$ is the univariate Student's t distribution function with $\nu + 1$ degrees of freedom. For a derivation of (2.12) see Demarta and McNeil (2005).

Fig. 2.2 `Tail dependence`: Illustration of upper and lower tail dependence: upper tail dependence (squares), lower tail dependence (triangles)

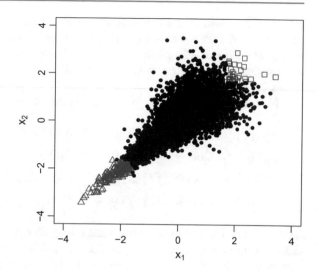

In contrast, some copulas have no tail dependence at all, neither upper nor lower tail dependence.

Ex 2.6 (Tail dependence coefficients for the bivariate Gaussian copula) For the bivariate Gaussian copula with parameter $\rho \in (-1, 1)$ (see Example 1.11 the *upper and lower tail dependence coefficient* are the same and are given by

$$\lambda = \lim_{x \to \infty} 2 \left(1 - \Phi \left(\frac{x\sqrt{1-\rho}}{\sqrt{1+\rho}} \right) \right) = 0.$$

Thus, the Gaussian copula has no tail dependence. See Demarta and McNeil (2005) and the references therein for details.

In practice, it is difficult to obtain a stable estimate the tail dependence coefficients in smaller sample sizes. Frahm et al. (2005) discuss estimation methods and pitfalls they encountered. Therefore simpler tail-weighted dependence measures have been proposed more recently. In particular, Krupskii and Joe (2015b) considered such measures. For a first exploration if tail dependence and in particular asymmetric tail dependence is present in the data *semi-correlations* are considered. Semi-correlations are derived from the correlation between transformations of the random variables X_1 and X_2 given by

$$\rho_N(C) := Cor(\Phi^{-1}(F_1(X_1)), \Phi^{-1}(F_2(X_2))),$$

where F_j are the marginal distribution functions of X_j for $j = 1, 2$. Here C is the associated copula to (X_1, X_2). The quantities $\Phi^{-1}(F_i(X_i))$ are called the *van der Waerden scores*, which were first defined in Van der Waerden (1953). Note that the van der Waerden scores are standard normally distributed. When instead of the marginal distribution functions the empirical distribution is used then one speaks of the van der Waerden score rank correlation discussed in Section III.6.1 of Hajek and Sidak (1967).

The corresponding semi-correlations are then defined as

$$\rho_N^{++}(C) := Cor(\Phi^{-1}(F_1(X_1)), \Phi^{-1}(F_2(X_2)|F_1(X_1) > .5, F_2(X_2) > .5)$$
$$\rho_N^{-+}(C) := Cor(\Phi^{-1}(F_1(X_1)), \Phi^{-1}(F_2(X_2)|F_1(X_1) < .5, F_2(X_2) > .5)$$
$$\rho_N^{+-}(C) := Cor(\Phi^{-1}(F_1(X_1)), \Phi^{-1}(F_2(X_2)|F_1(X_1) > .5, F_2(X_2) < .5)$$
$$\rho_N^{--}(C) := Cor(\Phi^{-1}(F_1(X_1)), \Phi^{-1}(F_2(X_2)|F_1(X_1) < .5, F_2(X_2) < .5).$$

These population semi-correlations can easily be estimated based on the observed data. Large values indicate the presence of tail dependence in the corresponding corner of the bivariate copula space $[0, 1] \times [0, 1]$. If the estimates of ρ_N^{++} and ρ_N^{--} differ considerably, then asymmetric tail dependence is present. For the bivariate normal distribution explicit expressions for the semi-correlations can be derived.

2.4 Partial and Conditional Correlations

In the case of d variables, we consider the dependence of any pair of variables. Additionally, we are interested in the dependence of two variables after the effect of the remaining variables are removed (partial correlations) or the dependence when we fix the values of the remaining variables (conditional correlations).

Definition 2.13 (*Partial regression coefficients and partial correlation*) Let $X_1, ...X_d$ be random variables with mean zero and variance σ_i^2. Further denote by $I_{-(i,j)}^d$ the set $\{1, ..., d\}$ with indices i and j for $i \neq j$ removed. Define the *partial regression coefficients* $b_{i,j;I_{-(i,j)}^d}$ for $i < j$ as quantities which minimize

$$E([X_i - \sum_{j=2, j \neq i}^{d} a_{i,j;I_{-(i,j)}^d} X_j]^2). \tag{2.13}$$

The corresponding $\binom{n}{2}$ *partial correlations* $\rho_{i,j;I_{-(i,j)}^d}$ are defined as

$$\rho_{i,j;I_{-(i,j)}^d} = \text{sgn}(b_{i,j;I_{-(i,j)}^d}) \times \sqrt{b_{i,j;I_{-(i,j)}^d} \times b_{j,i;I_{-(i,j)}^d}}. \tag{2.14}$$

The partial correlation can be interpreted as the correlation between the projection of X_i and X_j onto the plane orthogonal to the space spanned by $X_{-(i,j)}$. Here $X_{-(i,j)}$ is the vector of variables, where the variables X_i and X_j are removed. For the calculation of the partial correlations, Yule and Kendall (1950) showed the following *recursive formula*.

Theorem 2.14 (Recursion for partial correlations) *The partial correlations defined in* (2.14) *satisfy the following recursions:*

$$\rho_{i,j;I^d_{-(i,j)}} = \frac{\rho_{i,j;I^{d-1}_{-(i,j)}} - \rho_{i,d;I^{d-1}_{-(i,j)}}\,\rho_{j,d;I^{d-1}_{-(i,j)}}}{\sqrt{1 - \rho^2_{i,d;I^{d-1}_{-(i,j)}}}\sqrt{1 - \rho^2_{j,d;I^{d-1}_{-(i,j)}}}}. \tag{2.15}$$

Ex 2.7 (*Partial correlations in three dimensions*) This example is taken from page 69 of Kurowicka and Cooke (2006). For X_1, X_2, X_3 with zero mean and finite variance σ_i^2 the partial regression coefficients $b_{12;3}, b_{13;2}, b_{21;3}$, and $b_{23;1}$ minimize

$$\mathrm{E}([X_1 - a_{12;3}X_2 - a_{13;2}X_3]^2)$$
$$\mathrm{E}([X_2 - a_{21;3}X_1 - a_{23;1}X_3]^2)$$

over real valued $a_{12;3}, a_{13;2}, a_{21;3}$ and $a_{23;1}$. Taking partial derivatives and setting the above equations to zero the partial regression coefficients have to satisfy

$$\frac{\partial}{\partial b_{12;3}}\mathrm{E}([X_1 - b_{12;3}X_2 - b_{13;2}X_3]^2) = -2\mathrm{E}([X_1 - b_{12;3}X_2 - b_{13;2}X_3] \times X_2) = 0$$

$$\frac{\partial}{\partial b_{13;2}}\mathrm{E}([X_1 - b_{12;3}X_2 - b_{13;2}X_3]^2) = -2\mathrm{E}([X_1 - b_{12;3}X_2 - b_{13;2}X_3] \times X_3) = 0$$

$$\frac{\partial}{\partial b_{21;3}}\mathrm{E}([X_2 - b_{21;3}X_1 - b_{23;1}X_3]^2) = -2\mathrm{E}([X_2 - b_{21;3}X_1 - b_{23;1}X_3] \times X_2) = 0$$

$$\frac{\partial}{\partial b_{23;1}}\mathrm{E}([X_2 - b_{21;3}X_1 - b_{23;1}X_3]^2) = -2\mathrm{E}([X_2 - b_{21;3}X_1 - b_{23;1}X_3] \times X_3) = 0$$

This implies using the zero-mean assumption that

$$\mathrm{Cov}(X_1, X_2) - b_{12;3}\mathrm{Var}(X_2) - b_{13;2}\mathrm{Cov}(X_2, X_3) = 0$$
$$\mathrm{Cov}(X_1, X_3) - b_{12;3}\mathrm{Cov}(X_2, X_3) - b_{13;2}\mathrm{Var}(X_3) = 0 \tag{2.16}$$
$$\mathrm{Cov}(X_1, X_2) - b_{21;3}\mathrm{Var}(X_1) - b_{23;1}\mathrm{Cov}(X_1, X_3) = 0$$
$$\mathrm{Cov}(X_2, X_3) - b_{21;3}\mathrm{Cov}(X_1, X_3) - b_{23;1}\mathrm{Var}(X_3) = 0.$$

In the following, we shorten notation by setting $c_{ij} := \mathrm{Cov}(X_i, X_j)$ and $v_i := \mathrm{Var}(X_i)$. Solving for $b_{13;2}$ and $b_{23;1}$ using the second and fourth equation of (2.16) gives

$$b_{13;2} = \frac{c_{13} - b_{12;3}c_{23}}{v_3} \quad \text{and } b_{23;1} = \frac{c_{23} - b_{21;3}c_{13}}{v_3}.$$

Substituting these back into the first and third equation of (2.16) yields

$$b_{12;3} = \frac{\frac{c_{13} \times c_{23}}{v_3} - c_{12}}{\frac{c_{23}^2}{v_3} - v_2} \quad \text{and } b_{21;3} = \frac{\frac{c_{13} \times c_{23}}{v_3} - c_{12}}{\frac{c_{13}^2}{v_3} - v_1}.$$

Thus we obtain

$$\rho_{12;3} = \sqrt{b_{12;3} \times b_{21;3}}$$

$$= \sqrt{\frac{\frac{c_{13} \times c_{23}}{v_3} - c_{12}}{\frac{c_{23}^2}{v_3} - v_2} \times \frac{\frac{c_{13} \times c_{23}}{v_3} - c_{12}}{\frac{c_{13}^2}{v_3} - v_1}}$$

$$= \sqrt{\frac{\frac{c_{13} \times c_{23}}{v_3 v_2} - \frac{c_{12}}{v_2}}{\frac{c_{23}^2}{v_3 v_2} - \frac{v_2}{v_2}} \times \frac{\frac{c_{13} \times c_{23}}{v_3 v_1} - \frac{c_{12}}{v_1}}{\frac{c_{13}^2}{v_3 v_1} - \frac{v_1}{v_1}}}$$

$$= \sqrt{\frac{\frac{c_{13}^2 c_{23}^2}{v_3^2 v_1 v_2} - \frac{2 c_{12} c_{23} c_{13}}{v_3 v_1 v_2} + \frac{c_{12}^2}{v_2 v_1}}{(1 - \rho_{23}^2)(1 - \rho_{13}^2)}}$$

$$= \sqrt{\frac{\rho_{13}^2 \rho_{23}^2 - 2\rho_{12}\rho_{23}\rho_{13} + \rho_{12}^2}{(1 - \rho_{23}^2)(1 - \rho_{13}^2)}}$$

$$= \sqrt{\frac{(\rho_{12} - \rho_{13}\rho_{23})^2}{(1 - \rho_{23}^2)(1 - \rho_{13}^2)}}$$

$$= \frac{(\rho_{12} - \rho_{13}\rho_{23})}{\sqrt{(1 - \rho_{23}^2)(1 - \rho_{13}^2)}}.$$

This shows that the recursion (2.15) is valid for $d = 3$.

We now consider *conditional correlations*. Formally they are defined as the Pearson correlation of the bivariate distribution (X_i, X_j) given $X_{I^d_{-(i,j)}}$.

Table 2.2 WINE3: Estimated partial correlations for the red wine data of Example 1.9

$\hat{\rho}_{acf,acv;acc}$	$\hat{\rho}_{acf,acc;acv}$	$\hat{\rho}_{acv,acc;acf}$
0.19	0.66	−0.53

Definition 2.15 (*Conditional correlations*) The conditional correlation of (X_i, X_j) given $X_{-(i,j)}$ is defined as

$$
\rho_{ij|I^d_{-(i,j)}} := \frac{E(X_i X_j | X_{-(i,j)}) - E(X_i | X_{-(i,j)}) E(X_j | X_{-(i,j)})}{\sigma(X_i | X_{-(i,j)}) \sigma(X_j | X_{-(i,j)})},
$$

where $\sigma^2(X_i | X_{-(i,j)})$ and $\sigma^2(X_j | X_{-(i,j)})$ are the corresponding conditional variances.

Remark 2.16 (*Properties of conditional correlations*)

- For arbitrary bivariate distributions partial and conditional correlations are not the same in general, however in the class of elliptical distributions they coincide (see Baba et al. 2004).
- In general, the bivariate conditional distribution of (X_i, X_j) given $X_{-(i,j)}$ depends on the conditioning values, thus the associated moments also depend on those. However, under joint normality there is no dependence on the conditioning values.

Ex 2.8 (*Empirical partial and conditional correlation of the wine data of Example 1.9*) The estimated partial correlations associated with the data of Example 1.9 are given in Table 2.2. This shows that there is a strong partial correlation of acf and acc given acv. If we assume normality of the data, then the estimated partial correlations are also estimates of the conditional correlations.

2.5 Exercises

Exer 2.1

Tail dependence coefficients for the Student's t-distribution: Derive the expression (2.12) for the tail dependence coefficient of a bivariate Student's t distribution.

Table 2.3 WINE7: Estimated pairwise dependence measures (top: Kendall's τ, middle: Pearson correlations, bottom: Spearman's ρ_s)

Estimated pairwise Kendall's τ

	acf	acv	acc	clor	st	den	ph
acf	1.00	−0.19	0.48	0.18	−0.06	0.46	−0.53
acv	−0.19	1.00	−0.43	0.11	0.06	0.02	0.16
acc	0.48	−0.43	1.00	0.08	0.01	0.25	−0.39
clor	0.18	0.11	0.08	1.00	0.09	0.29	−0.16
st	−0.06	0.06	0.01	0.09	1.00	0.09	−0.01
den	0.46	0.02	0.25	0.29	0.09	1.00	−0.22
ph	−0.53	0.16	−0.39	−0.16	−0.01	−0.22	1.00

Estimated pairwise Pearson correlations ρ

	acf	acv	acc	clor	st	den	ph
acf	1.00	−0.26	0.67	0.09	−0.11	0.67	−0.68
acv	−0.26	1.00	−0.55	0.06	0.08	0.02	0.23
acc	0.67	−0.55	1.00	0.20	0.04	0.36	−0.54
clor	0.09	0.06	0.20	1.00	0.05	0.20	−0.27
st	−0.11	0.08	0.04	0.05	1.00	0.07	−0.07
den	0.67	0.02	0.36	0.20	0.07	1.00	−0.34
ph	−0.68	0.23	−0.54	−0.27	−0.07	−0.34	1.00

Estimated pairwise Spearman's ρ_s

	acf	acv	acc	clor	st	den	ph
acf	1.00	−0.28	0.66	0.25	−0.09	0.62	−0.71
acv	−0.28	1.00	−0.61	0.16	0.09	0.03	0.23
acc	0.66	−0.61	1.00	0.11	0.01	0.35	−0.55
clor	0.25	0.16	0.11	1.00	0.13	0.41	−0.23
st	−0.09	0.09	0.01	0.13	1.00	0.13	−0.01
den	0.62	0.03	0.35	0.41	0.13	1.00	−0.31
ph	−0.71	0.23	−0.55	−0.23	−0.01	−0.31	1.00

Exer 2.2

URAN3: *Three-dimensional uranium data:* Perform the same analysis as for the wine data in Example 2.8 for variables Co, Ti, and Sc of the data uranium contained in the R package copula of Hofert et al. (2017). This data was also considered in Acar et al. (2012) and Kraus and Czado (2017b).

Exer 2.3

ABALONE3: *Three-dimensional abalone data:* Perform the same analysis as for the wine data in Example 2.8 for variables shucked, viscera and shell of the data `abalone` contained in the R package `PivotalR` from Pivotal Inc. (2017). Again restrict your analysis to the female abalone shells.

Exer 2.4

WINE7: *Estimated bivariate dependencies:* Consider again the data of Exercise 1.7. Estimates of bivariate dependence measures are given in Table 2.3. Interpret the strength between the pairs of variables.

Exer 2.5

Calculation of partial correlations: Show that the conditional correlation $\rho_{ij|I^d_{-(i,j)}}$ when $X = (X_1, ..., X_d) \sim N_d(\mu, \Sigma)$ can be calculated as the (i, j)th element of the matrix Σ^{-1}.

Exer 2.6

Partial correlation is equal to conditional correlation under normality for $d = 3$: Show that the partial correlation $\rho_{12;3}$ is equal to the conditional correlation $\rho_{12|3}$, when the random variables (X_1, X_2, X_3) are jointly normal distributed.

Exer 2.7

Connection between partial and ordinary correlations: Consider the following set of partial and ordinary correlations $S_p := \{\rho_{12}, \rho_{23}, \rho_{34}, \rho_{13;2}, \rho_{24;3}, \rho_{14;23}\}$ and the set of ordinary correlations $S := \{\rho_{12}, \rho_{23}, \rho_{34}, \rho_{13}, \rho_{24}, \rho_{14}\}$ between four random variables. Using Theorem 2.14 show that there exist a one-to-one relationship between the sets S_p and S.

Bivariate Copula Classes, Their Visualization, and Estimation

3

3.1 Construction of Bivariate Copula Classes

There are three major construction approaches of copulas. One arising from applying the probability integral transform (see Definition 1.3) to each margin of known multivariate distributions and one to use generator functions. The first approach applied to elliptical distributions yields the class of *elliptical copulas*. With the second approach, we obtain the class of *Archimedean copulas*. The well-known examples of this class are the Clayton, Gumbel, Frank, and Joe copula families. The third approach arises from extensions of univariate extreme-value theory to higher dimensions.

3.2 Bivariate Elliptical Copulas

In Chap. 1, we have already seen two members of the class of elliptical copulas, namely, the multivariate Gauss copula (Example 1.12) and the multivariate Student's t copula (Example 1.14). Here we consider the bivariate versions (Examples 1.11 and 1.13).

3.3 Archimedean Copulas

Archimedean copulas are discussed extensively in the books by Joe (1997) and Nelsen (2006). A more recent characterization of the generator functions can be found in McNeil and Nešlehová (2009). Here we give an elementary introduction to bivariate Archimedean copulas.

© Springer Nature Switzerland AG 2019
C. Czado, *Analyzing Dependent Data with Vine Copulas*, Lecture Notes in Statistics 222, https://doi.org/10.1007/978-3-030-13785-4_3

Definition 3.1 (*Bivariate Archimedean copulas*) Let Ω be the set of all continuous, strictly monotone decreasing, and convex functions $\varphi : I \to [0, \infty]$ with $\varphi(1) = 0$. Let $\varphi \in \Omega$, then

$$C(u_1, u_2) = \varphi^{[-1]}(\varphi(u_1) + \varphi(u_2)) \qquad (3.1)$$

is a copula. C is called a *bivariate Archimedean copula* with *generator* φ. Is $\varphi(0) = \infty$, the generator is called *strict*. Here $\varphi^{[-1]}$ is the pseudo-inverse of φ, which is defined as $\varphi^{[-1]} : [0, \infty] \to [0, 1]$ with

$$\varphi^{[-1]}(t) := \begin{cases} \varphi^{-1}(t) & , 0 \le t \le \varphi(0) \\ 0 & , \varphi(0) \le t \le \infty. \end{cases}$$

Remark 3.2 (*Properties of generators of Archimedean copulas*)

1. If $\varphi(0) = \infty$, then $\varphi^{[-1]}(t) = \varphi^{-1}(t)$ $\forall t \in [0, \infty]$.
2. The pseudo-inverse $\varphi^{[-1]}$ is continuous, nonincreasing on $[0, \infty]$ and strictly decreasing on $[0, \varphi(0)]$.
3. It satisfies $\varphi^{[-1]}(\varphi(u)) = u$ on $[0, 1]$ and

$$\varphi(\varphi^{[-1]}(t)) = \begin{cases} t & , 0 \le t \le \varphi(0) \\ \varphi(0) & , \varphi(0) \le t \le \infty. \end{cases}$$

4. Some authors prefer to express Archimedean copulas in terms of $\psi(w) := \varphi^{[-1]}(w)$ for $w > 0$. This allows for simpler expressions for associated stochastic representations.

Lemma 3.3 (Density of Archimedean copulas) *If C is a continuous Archimedean copula, then its density c can be expressed as*

$$c(u_1, u_2) = \frac{\partial^2 C(u_1, u_2)}{\partial u_1 \partial u_2} = \frac{\varphi''(C(u_1, u_2))\varphi'(u_1)\varphi'(u_2)}{[\varphi'(C(u_1, u_2))]^3}. \qquad (3.2)$$

We discuss now examples of *parametric bivariate Archimedean copulas* with a single parameter.

Ex 3.1 (*Parametric bivariate Archimedean copulas with a single parameter*)
The *bivariate Clayton copula* is given as

$$\textbf{Clayton:}\quad C(u_1, u_2) = (u_1^{-\delta} + u_2^{-\delta} - 1)^{-\frac{1}{\delta}}, \tag{3.3}$$

where $0 < \delta < \infty$ control the degree of dependence. Full dependence is
obtained when $\delta \to \infty$. Independence corresponds to $\delta \to 0$.
   The *bivariate Gumbel copula* is given as

$$\textbf{Gumbel:}\quad C(u_1, u_2) = \exp[-\{(-\ln u_1)^{\delta} + (-\ln u_2)^{\delta}\}^{\frac{1}{\delta}}], \tag{3.4}$$

where $\delta \geq 1$ is the parameter of dependence. For $\delta \to \infty$ we have full dependence, while $\delta = 1$ corresponds to independence.
   The *bivariate Frank copula* is given as

$$\textbf{Frank:}\quad C(u_1, u_2) = -\frac{1}{\delta} \ln\left(\frac{1}{1 - e^{-\delta}}[(1 - e^{-\delta}) - (1 - e^{-\delta u_1})(1 - e^{-\delta u_2})]\right), \tag{3.5}$$

where the parameter δ can take values in $[-\infty, \infty]\backslash\{0\}$. For $\delta \to 0^+$, the
independence copula arises.
   The *bivariate Joe copula* is defined as

$$\textbf{Joe:}\quad C(u_1, u_2) = 1 - \left((1 - u_1)^{\delta} + (1 - u_2)^{\delta} - (1 - u_1)^{\delta}(1 - u_2)^{\delta}\right)^{\frac{1}{\delta}}, \tag{3.6}$$

where $\delta \geq 1$. The independence copula corresponds to $\delta = 1$. The copula
densities are illustrated in Fig. 3.1 where the copula parameter of each of the
four Archimedean copulas results in a Kendall's $\tau = .7$. In particular, this occurs
if $\delta = 4.7$ for the Clayton copula, $\delta = 3.3$ for the Gumbel copula, $\delta = 11$ for
the Frank copula, and $\delta = 5.5$ for the Joe copula. For details on the relationship
between the parameters of Archimedean copulas and associated Kendall's τ see
Sect. 3.5.

There are also Archimedean copulas available with two parameters such as the
BB1 and *BB7* families.

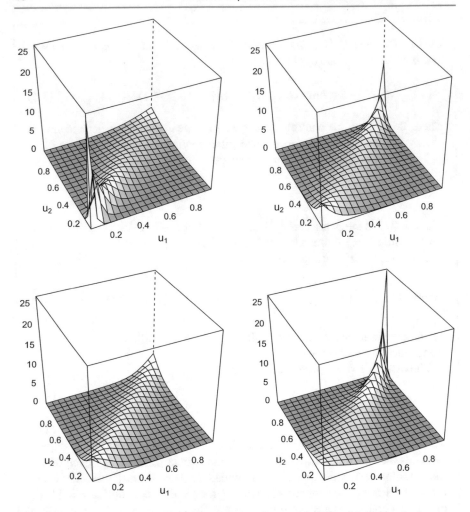

Fig. 3.1 `Archimedean copulas`: Bivariate Archimedean copula densities with a single parameter: top left: Clayton, top right: Gumbel, bottom left: Frank, and bottom right: Joe (the copula parameter is chosen in such a way that the corresponding Kendall's $\tau = .7$)

Ex 3.2 (BB copulas) The BB notation was introduced in Joe (1997), and the properties of several members of the BB family were explored in Nikoloulopoulos et al. (2012). A summary of the properties of the BB family is also contained in Joe (2014). The *bivariate BB1 copula* is defined as

$$\textbf{BB1}: \ C(u, v; \theta, \delta) = \left\{ 1 + [(u^{-\theta} - 1)^{\delta} + (v^{-\theta} - 1)^{\delta}]^{\frac{1}{\delta}} \right\}^{-\frac{1}{\theta}}$$
$$= \eta(\eta^{-1}(u) + \eta^{-1}(v)), \tag{3.7}$$

where $\eta(s) = \eta_{\theta,\delta}(s) = (1 + s^{\frac{1}{\delta}})^{-\frac{1}{\theta}}$. Note that η corresponds to the inverse generator. Here the parameter range is $\theta > 0$ and $\delta \geq 1$. For $\theta \to 0^+$ and the $\delta \to 1^+$, the independence copula arises. The *bivariate BB7 copula* is defined as

BB7:
$$C(u, v; \theta, \delta) = 1 - \left(1 - [(1 - (1 - u)^\theta)^{-\delta} + (1 - (1 - v)^\theta)^{-\delta} - 1]^{-\frac{1}{\delta}}\right)^{\frac{1}{\theta}}$$
$$= \eta(\eta^{-1}(u) + \eta^{-1}(v)),$$
(3.8)

where $\eta(s) = \eta_{\theta,\delta}(s) = 1 - [1 - (1 + s)^{-\frac{1}{\delta}}]^{\frac{1}{\theta}}$, $\theta \geq 1$ and $\delta > 0$. The independence copula corresponds to $\theta = 1$ and $\delta = 0$. We now illustrate the corresponding copula densities. Since both families have two parameters, there are more than one parameter set to achieve a fixed Kendall'τ. For both families, we use two parameter sets for a resulting $\tau = .7$ and the corresponding copula density plots are shown in Fig. 3.2.

3.4 Bivariate Extreme-Value Copulas

In extreme-value theory, one is interested in studying the behavior of extreme events and their dependence. The foundation for probabilistic modeling of multivariate extremes including asymptotic theory was developed in de Haan and Resnick (1977) and Pickands (1981). Books which focus on the statistical inference for extreme-value distributions are Coles et al. (2001), Beirlant et al. (2006) and McNeil et al. (2015). The connection of multivariate extreme-value theory to copulas has been investigated in Gudendorf and Segers (2010). Genest and Nešlehová (2013) give a survey of using copulas to model extremes. We restrict in this short exposition to the bivariate case. In particular, we consider n i.i.d. bivariate random vectors $X_i = (X_{i1}, X_{i2})^\top$ for $i = 1, \ldots, n$ with common bivariate distribution function F and margins F_1, F_2. Further, let C_X be the corresponding copula to F. In multivariate extreme-value theory often the behavior of component-wise maxima defined by

$$M_{nj} = \max_{i=1,\ldots,n} (X_{ij}) \text{ for } j = 1, 2$$
(3.9)

is studied. It is easy to show that the copula C_{M_n} associated with $\mathbf{M}_n = (M_{n1}, M_{n2})^\top$ is given by

$$C_{M_n}(u_1, u_2) = [C_X(u_1^{\frac{1}{n}}, u_2^{\frac{1}{n}})]^n$$
(3.10)

Even though the vector of component-wise maxima does not in general constitute a sample point, multivariate extreme-value theory studies the limiting behavior of (3.10).

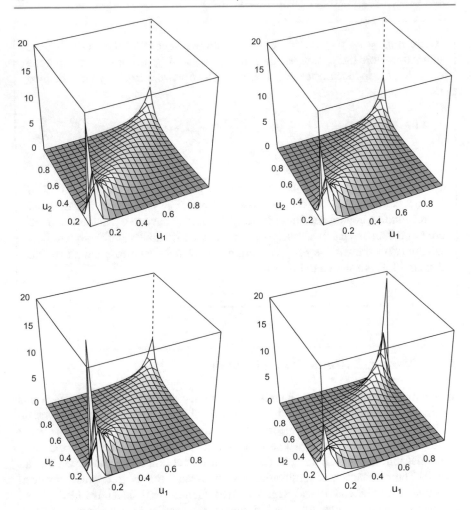

Fig. 3.2 BB copula densities: Top left: BB1 with $\delta = 20/12$, $\theta = 2$, top right: BB1 with $\delta = 2$, $\theta = 4/3$, bottom left: BB7 with $\theta = 2$, $\delta = 4.5$, and bottom right: BB7 with $\theta = 1.3$, $\delta = 2$ (copula parameters are chosen in such a way that the corresponding Kendall's $\tau = .7$)

Definition 3.4 (*Bivariate extreme-value copula*) A bivariate copula C is called an *extreme-value copula* if there exists a bivariate copula C_X such that for $n \to \infty$ we have

$$[C_X(u_1^{1/n}, u_2^{1/n})]^n \to C(u_1, u_2) \forall (u_1, u_2) \in [0, 1]^2. \quad (3.11)$$

The copula C_X is said to be in the domain of attraction of the copula C.

One characterization of bivariate extreme-value copulas involves *max stable copulas*.

Definition 3.5 A bivariate copula C is called *max stable* if it satisfies

$$C(u_1, u_2) = [C(u_1^{1/m}, u_2^{1/m})]^m \tag{3.12}$$

for every integer $m \geq 1$ and for all $(u_1, u_2) \in [0, 1]^2$.

This definition allows in view of (3.11) the following characterization.

Theorem 3.6 (Max stability and extreme-value copula) *A bivariate copula C is an extreme-value copula if and only if it is a max stable copula.*

The proof is left as an exercise. So far, the definition of an extreme-value and a max stable copula can be easily extended to the d-dimensional case and also Theorem 3.6 remains valid. It is now possible to characterize a d-dimensional extreme-value copula in terms of a stable tail dependence function. To define the stable tail dependence function one requires the notion of a spectral measure. Details can be found, for example, in Beirlant et al. (2006).

The following characterization of extreme value copulas in terms of the *Pickands dependence function A* is only valid in the bivariate case.

Theorem 3.7 (Characterization of bivariate extreme-value copulas in terms of the Pickands dependence function) *A bivariate copula C is an extreme-value copula if and only if*

$$C(u_1, u_2) = \exp\left\{ [\ln(u_1) + \ln(u_2)] A\left(\frac{\ln(u_2)}{\ln(u_1 u_2)} \right) \right\}, \tag{3.13}$$

where the function $A : [0, 1] \to [\frac{1}{2}, 1]$ is convex and satisfies $\max\{1 - t, t\} \leq A(t) \leq 1$ for all $t \in [0, 1]$. The function A is called the Pickands dependence function.

The proof starts with the representation of extreme-value copulas in terms of their stable tail dependence function restricted to two dimensions and uses the unit simplex $\{(1 - t, t), t \in (0, 10\}$ to express the stable tail dependence function in terms of the Pickands dependence function. An illustration of the admissible range of the Pickands dependence function is given in Fig. 3.3.

Fig. 3.3 Pickands
dependence function:
The admissible range is
colored in gray. The dashed
upper line corresponds to the
independence copula while
the solid line denotes
max{t, $1 - t$}, which
corresponds to perfectly
dependent variables. The
dotted lines show two
examples of typical
dependence functions

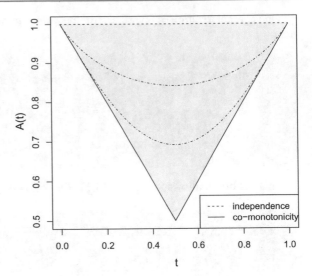

Using Theorem 3.7, it is easy to express the conditional distribution functions
and the copula density of bivariate extreme-value copulas in terms of their Pickands
dependence function.

Lemma 3.8 (Conditional distribution functions and copula density of a bivariate extreme-value copula) *Assuming that the Pickands dependence function A
of the bivariate extreme-value copula C is twice differentiable the conditional
distributions and the copula density can be expressed as follows:*

$$C_{2|1}(u_2|u_1) = \frac{C(u_1, u_2)}{u_1}(A(t) - tA'(t)) \qquad (3.14)$$

$$C_{1|2}(u_1|u_2) = \frac{C(u_1, u_2)}{u_2}(A(t) + (1 - t)A'(t)) \qquad (3.15)$$

$$c(u_1, u_2) = \frac{C(u_1, u_2)}{u_1 u_2}$$
$$\cdot \left[A(t)^2 + (1 - 2t)A'(t)A(t) - (1 - t)t \left(A'(t)^2 - \frac{A''(t)}{\ln(u_1 u_2)} \right) \right] \qquad (3.16)$$

where $t = \ln(u_2)/\ln(u_1 u_2)$.

There are many parametric families of bivariate extreme-value copulas known in
the literature. One special bivariate extreme-value copula is the *Marshall–Olkin copula* (Marshall and Olkin 1967). It is the copula associated with dependent lifetimes
of two components, which fail after the occurrence of a shock. Shocks can occur to a

single component or jointly to both components. For more details see, for example, Durante et al. (2015). Mai and Scherer (2010) showed that the Marshall–Olkin copula is max stable and thus an extreme-value copula. Its Pickands dependence function is not continuous and thus statistical inference is more difficult to derive. However, the Marshall–Olkin copula occurs often as limiting case of other extreme-value copulas.

The next example of a parametric extreme-value copula is the *Hüsler–Reiss copula* with parameter $\lambda \geq 0$. It arises as the associated copula from taking block maxima of a bivariate normal distribution, where the correlation parameter converges to one in addition to the block size going to infinity. In particular, this means that the correlation parameter ρ depends on the block size n, i.e., $\rho = \rho_n$, and it is assumed that $\rho_n \to 1$ for $n \to \infty$. Additionally, one requires that $(1 - \rho_n)\ln(n) \to \lambda^2$ for a $\lambda \geq 0$ for $n \to \infty$. This was shown in Hüsler and Reiss (1989). Similarly, the *t extreme-value (t-EV) copula* arises from the bivariate Student's t distribution. A derivation of the t-EV copula is given in Demarta and McNeil (2005).

The only Archimedean copula, which is also extreme value, is the Gumbel copula. It is obtained as the limit of block maxima of bivariate Archimedean copulas with generator φ. More precisely, the generator has to be differentiable and the following limit has to exist (Gudendorf and Segers 2010):

$$\theta := -\lim_{s\downarrow 0} \frac{s\varphi'(1 - s)}{\varphi(1 - s)} \in [1, \infty].$$

In this case the Gumbel copula with parameter θ as the limit of block maxima arises. If one considers the survival copula (for precise definition see Example 3.3) of an Archimedean copula with differentiable generator φ, Gudendorf and Segers (2010) note the following: If the limit

$$\delta := -\lim_{s\downarrow 0} \frac{\varphi(s)}{\varphi'(s)} \in [0, \infty]$$

exist, the Galambos copula with parameter δ first discussed in Galambos (1975) arises. Tawn (1988) introduced a three parameter extreme-value copula family, which allows for very flexible shapes (see also Fig. 3.10).

Table 3.1 is taken from Eschenburg (2013) and gives the Pickands dependence function of these common extreme-value copulas together with their domain of attraction.

Figure 3.4 also taken from Eschenburg (2013) displays the interconnectedness of the bivariate extreme-value copulas given in Table 3.1. In particular the Gumbel copula can be obtained from the Tawn copula (with parameters $\psi_1 = \psi_2 = 1$) as well as from the BB5 copula (with parameter $\delta \to 0$). Similarly, the Galambos copula can be obtained from the extended Joe copula (with parameters $\psi_1 = \psi_2 = 1$) as well as from the BB5 copula (with parameter $\theta = 1$). The extended Joe copula with $\psi_1 = \psi_2 = 1$ is the standard Joe copula discussed, for example, on page 170 of Joe (2014).

Table 3.1 Overview: Extreme-value copula families and their Pickands dependence functions (Eschenburg 2013)

Copula family	Pickands dependence function	Parameter(s)	Domain of attraction	Symmetry
Marshall–Olkin	$A(t) = \max\{1 - \alpha_1(1-t), 1 - \alpha_2 t\}$	$0 \leq \alpha_1, \alpha_2 \leq 1$		No
Hüsler–Reiss	$A(t) = (1-t)\Phi(z_{1-t}) + t\Phi(z_t)$, $z_t = (\frac{1}{\lambda} + \frac{\lambda}{2}\ln\frac{t}{1-t})$	$\lambda \geq 0$	Gaussian copula	Yes
t-EV	$A(t) = (1-t) \cdot T_{\nu+1}(z_{1-t}) + t \cdot T_{\nu+1}(z_t)$, $z_t = (1 + \nu)^{1/2}\left(\left[\frac{t}{1-t}\right]^{1/\nu} - \rho\right)(1-\rho^2)^{-1/2}$	$\nu > 0$, $-1 < \rho < 1$	Student's t copula	Yes
Gumbel	$A(t) = [t^\theta + (1-t)^\theta]^{1/\theta}$	$\theta \geq 1$	Archimedean copula	Yes
Tawn	$A(t) = (1-\psi_1)(1-t) + (1-\psi_2)t + [(\psi_1(1-t))^\theta + (\psi_2 t)^\theta]^{1/\theta}$	$0 \leq \psi_1, \psi_2 \leq 1$, $\theta \geq 1$		No
Galambos	$A(t) = 1 - [t^{-\delta} + (1-t)^{-\delta}]^{-1/\delta}$	$\delta > 0$	Archimedean survival copula	Yes
Extended Joe (BB8)	$A(t) = 1 - \{[\psi_1(1-t)]^{-\delta} + (\psi_2 t)^{-\delta}\}^{-1/\delta}$	$0 \leq \psi_1, \psi_2 \leq 1$, $\delta > 0$		No
BB5	$A(t) = \{t^\theta + (1-t)^\theta - [(1-t)^{-\theta\delta} + t^{-\theta\delta}]^{-1/\delta}\}^{1/\theta}$	$\theta \geq 1, \delta > 0$		Yes

Finally, we illustrate the bivariate Tawn copula density with three parameters for four choices in Fig. 3.5. In particular, the four parameter set choices are

$$\text{Set 1: } \theta = 2, \psi_1 = .8, \psi_2 = .2$$

$$\text{Set 2: } \theta = 2, \psi_1 = .8, \psi_2 = .8$$

$$\text{Set 3: } \theta = 5, \psi_1 = .8, \psi_2 = .2$$

$$\text{Set 4: } \theta = 5, \psi_1 = .8, \psi_2 = .8.$$

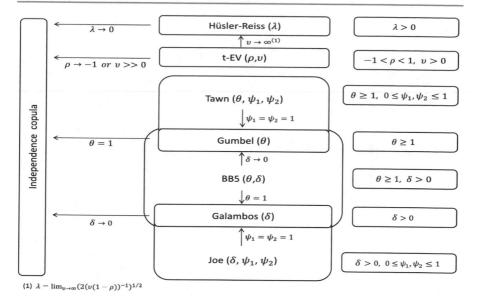

(1) $\lambda - \lim_{v \to \infty} (2(v(1-\rho))^{-1})^{1/2}$

Fig. **3.4** Relationships between the bivariate extreme-value copulas: Parameter ranges of the families are given on the right side. An arrow symbolizes that a family can be obtained from another family by the given parameter(s). Arrows to the left display convergence toward the independence copula (Eschenburg 2013)

3.5 Relationship Between Copula Parameters and Kendall's τ

For bivariate Archimedean, extreme-value, and elliptical copulas, there exist general results relating the Kendall's τ to generators, the Pickands dependence function or parameters of the copula families. In particular, the following result is available.

Theorem 3.9 (Kendall's τ for bivariate Archimedean, extreme-value, and elliptical copulas) *Let φ a generator of a bivariate Archimedean copula and A the Pickands dependence function of a bivariate extreme-value copula with existing first derivative. The corresponding Kendall's τ for the Archimedean copula satisfies*

$$\tau = 1 + 4 \int_0^1 \frac{\varphi(t)}{\varphi'(t)} dt \tag{3.17}$$

and for the bivariate extreme we have

$$\tau = \int_0^1 \frac{t(1-t)}{A(t)} dA'(t). \tag{3.18}$$

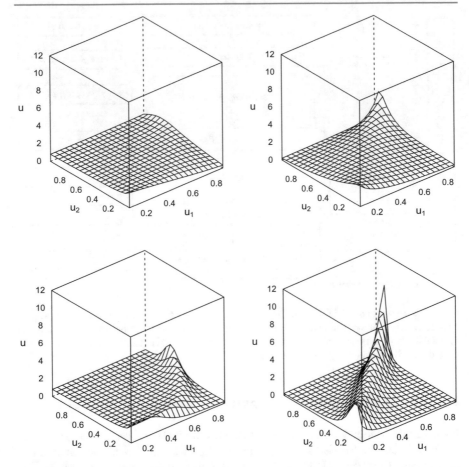

Fig. 3.5 Tawn copula densities: Top left: $\theta = 2, \psi_1 = .8, \psi_2 = .2$, top right: $\theta = 2$, $\psi_1 = .8, \psi_2 = .8$, bottom left: $\theta = 5, \psi_1 = .8, \psi_2 = .2$, and bottom right: $\theta = 5, \psi_1 = .8, \psi_2 = .8$

For elliptical copulas, we obtain the following relationship between the asso-ciation parameter ρ and Kendall's τ

$$\rho = \sin\left(\frac{\pi}{2}\tau\right).$$ (3.19)

The proof for elliptical copulas is given, for example, in Embrechts et al. (2003), while the proof for the Archimedean and the extreme-value copula can be found in Hürlimann (2003). In particular, we see that extreme-value copulas can only model positive dependence.

Table 3.2 Relationships: Kendall's τ and copula parameters for different bivariate copula families

Family	Kendall's τ	Range of τ
Gaussian	$\tau = \frac{2}{\pi}\arcsin(\rho)$	$[-1, 1]$
t	$\tau = \frac{2}{\pi}\arcsin(\rho)$	$[-1, 1]$
Gumbel	$\tau = 1 - \frac{1}{\delta}$	$[0, 1]$
Clayton	$\tau = \frac{\delta}{\delta+2}$	$[0, 1]$
Frank	$\tau = 1 - \frac{4}{\delta} + 4\frac{D_1(\delta)}{\delta}$ with $D_1(\delta) = \int_o^\delta \frac{x/\delta}{e^x - 1}dx$ (Debye function)	$[-1, 1]$
Joe	$\tau = 1 + \left(\frac{-2 + 2\gamma + 2\ln(2) + \Psi(\frac{1}{\delta}) + \Psi(\frac{1}{2}\frac{2+\delta}{\delta}) + \delta}{-2+\delta} \right)$ with Euler constant $\gamma = \lim_{n\to\infty}\left(\sum_{i=1}^n \frac{1}{i} - \ln(n) \right) \approx 0,57721$ and digamma function $\Psi(x) = \frac{d}{dx}\ln(\Gamma(x)) = \frac{d}{dx}\Gamma(x)/\Gamma(x)$	$[0, 1]$
BB1	$\tau = 1 - \frac{2}{\delta(\theta+2)}$	$[0, 1]$
BB7	$\tau = 1 - \frac{2}{\delta(2-\theta)} + \frac{4}{\theta^2\delta}B(\frac{2-2\theta}{\theta}+1, \delta+2)$ for $1 < \theta < 2$ with beta function $B(x, y) = \int_0^1 t^{x+1}(1-t)^{y-1}dt$	$[0, 1]$

Now we will summarize the relationships between Kendall's τ and the copula parameters in Table 3.2 for the discussed Archimedean and elliptical bivariate parametric copula families. For many of the bivariate extreme-value copula families given in Table 3.1, there is no simplification possible over the general expression given in (3.18).

We see from Table 3.2 that the Gumbel, Clayton, Joe, BB1, and BB7 copulas can only accommodate positive dependence ($\tau > 0$). We can extend this range by allowing for *rotations* which we will discuss in the next section.

In Fig. 2.1, we investigated the relationship of Kendall's τ and Spearman's ρ_s as the Pearson correlation parameter ρ changes in any bivariate elliptical copula. In Fig. 3.6, we show similar results for the Clayton and Gumbel copula. This shows that there is a considerable difference between the associated Kendall's τ and Spearman's ρ_s for the Clayton copula and somewhat less so for the Gumbel copula. In any case, the difference is higher compared to the ones observed in Fig. 2.1 for the elliptical copulas.

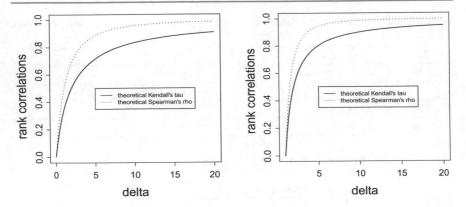

Fig. 3.6 Relationship: Copula parameter and rank-based dependence measures: Clayton copula (left) and Gumbel copula (right)

3.6 Rotated and Reflected Copulas

To extend the range of dependence we use counterclockwise rotations of the copula density $c(\cdot, \cdot)$ by

- $90°$: $c_{90}(u_1, u_2) := c(1 - u_2, u_1)$,
- $180°$: $c_{180}(u_1, u_2) := c(1 - u_1, 1 - u_2)$, and
- $270°$: $c_{270}(u_1, u_2) := c(u_2, 1 - u_1)$.

This allows, for example, to extend the Clayton copula to a copula with a full range of Kendall's τ values by defining

$$c_{clayton}^{extended}(u_1, u_2; \delta) := \begin{cases} c_{clayton}(u_1, u_2) & \text{if } \delta > 0 \\ c_{clayton}(1 - u_2, u_1) & \text{otherwise} \end{cases}$$

Here a $90°$ rotation is used for $\delta \leq 0$, but we also could use a $270°$ rotation. All rotations of the copula density are illustrated in Fig. 3.7 using normalized contour plots defined in Definition 3.11. The term rotation is used in the context of the copula density and does not correspond to rotations of the random vector (U_1, U_2).

Some rotations of bivariate copula density can also be explained by considering *reflections* of uniform random variables on $(0, 1)$. If (U_1, U_2) have copula C then the copula associated with the bivariate random vector $(1 - U_1, 1 - U_2)$ is the survival copula $\overline{C}$ discussed in Example 3.3 with copula density given by c_{180}. Recently, Joe (2014) considered on p. 272 further reflections such as $(1 - U_1, U_2)$ with the *1-reflected copula* $C^{*1}(u_1, u_2) = u_2 - C(1 - u_1, u_2)$ and density $c^{*1}(u_1, u_2) = c(1 - u_1, u_2)$ and $(U_1, 1 - U_2)$ with the *2-reflected copula* $C^{*2}(u_1, u_2) = u_1 - C(u_1, 1 - u_2)$ and density $c^{*2}(u_1, u_2) = c(u_1, 1 - u_2)$. For further discussion see Exercise 3.14.

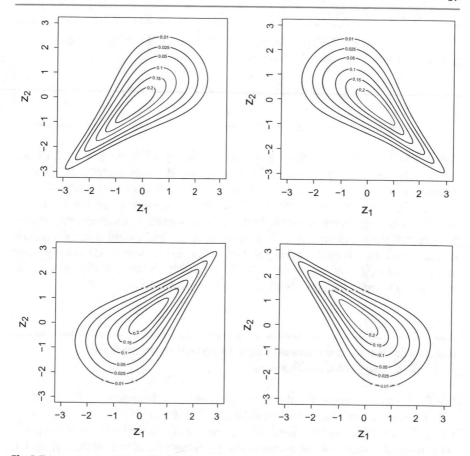

Fig. 3.7 Rotations: Normalized contour plots of Clayton rotations: top left: 0° rotation ($\tau = .5$), top right: 90° rotation ($\tau = -.5$), bottom left: 180° rotation ($\tau = .5$), and bottom right: 270° rotation ($\tau = -.5$)

Ex 3.3 (Bivariate survival copulas) In univariate survival analysis, one models the survival function $S(t) := P(T > t) = 1 - F(t)$ of a lifetime T instead of the distribution function $F(t)$. For bivariate survival times (T_1, T_2) the bivariate survival function $S(t_1, t_2) := P(T_1 > t_1, T_2 > t_2)$ can be expressed using a version of Sklar's theorem as

$$S(t_1, t_2) = \overline{C}(S_1(t_1), S_2(t_2)),$$

where $S_k(t_k)$ are the marginal survival functions of T_k for $k = 1, 2$. The copula $\overline{C}$ is called the *survival copula*. If we denote the associated copula to the lifetimes (T_1, T_2) by $C(u_1, u_2)$, we have the following relationship

$$\overline{C}(u_1, u_2) = u_1 + u_2 + C(1 - u_1, 1 - u_2) - 1 \qquad (3.20)$$

for $0 \leq u_1, u_2, \leq 1$. Using differentiation of (3.20), it is straightforward to see for the associated copula densities that

$$\bar{c}(u_1, u_2) = c(1 - u_1, 1 - u_2) = c_{180}(u_1, u_2).$$

Remark 3.10 (*Exchangeability and reflection symmetry of bivariate copulas*) Bivariate copulas which satisfy $c(u_1, u_2) = c(u_2, u_1)$ are called *exchangeable* or *reflection symmetric* around the $u_1 = u_2$ axis. The one parametric copulas considered so far are exchangeable. However, a 90 or 270° rotated version of the Clayton or Gumbel copula is no longer exchangeable. In the case of nonsymmetric bivariate copulas, the order of the arguments plays a role, so we might include them in the notation of the copula, i.e., we write $c_{12}(u_1, u_2)$ or $c_{21}(u_2, u_1)$. However, the density value of the random (U_1, U_2) is the same as of the random vector (U_2, U_1), so we have $c_{12}(u_1, u_2) = c_{21}(u_2, u_1)$ holds for every $(u_1, u_2) \in [0, 1]^2$.

3.7 Relationship Between Copula Parameters and Tail Dependence Coefficients

For the bivariate Gaussian and Student's t copula, we have already discussed the tail dependence coefficients (see Definition 2.12) in Examples 2.6 and 2.12, respectively. For the bivariate Archimedean copulas the tail dependence coefficients can be expressed in terms of the corresponding generator functions. Similarly, the tail dependence coefficients for bivariate extreme-value copulas can be written utilizing the Pickands dependence function. These results are summarized in Table 3.3. For details on the derivation see, for example, Embrechts et al. (2003) and Nikoloulopoulos et al. (2012) .

3.8 Exploratory Visualization

In this section, we are only considering bivariate parametric copulas, and therefore, we can study the shape of the copula itself or the copula density, when it exists. As in the univariate case, the exploration of the density is more insightful compared to its distribution function (copula). Since the support of the copula is the unit square, the copula densities for the different classes discussed before are not easy to interpret. Therefore, we consider the transformation to a bivariate distribution with normal $N(0, 1)$ margins and density $g(z_1, z_2)$. For example, if we consider a Gaussian copula, then this transformation just yields a bivariate normal distribution with zero mean vector, unit variances, and correlation ρ, i.e., contour plots of its density

Table 3.3 `Tail behavior:` Tail dependence coefficients of different bivariate copula families in terms of their copula parameters

Family	Upper tail dependence	Lower tail dependence
Gaussian	–	–
t	$2t_{\nu+1}\left(-\sqrt{\nu+1}\sqrt{\frac{1-\rho}{1+\rho}}\right)$	$2t_{\nu+1}\left(-\sqrt{\nu+1}\sqrt{\frac{1-\rho}{1+\rho}}\right)$
Gumbel	$2-2^{1/\delta}$	–
Clayton	–	$2^{-1/\delta}$
Frank	–	–
Joe	$2-2^{1/\delta}$	–
BB1	$2-2^{1/\delta}$	$2^{-1/(\delta\theta)}$
BB7	$2-2^{1/\theta}$	$2^{-1/\delta}$
Galambos	$2^{-1/\delta}$	–
BB5	$2-(2-2^{-1/\delta})^{1/\theta}$	–
Tawn	$(\psi_1+\psi_2)-(\psi_1^\theta+\psi_2^\theta)^{1/\theta}$	–
t-EV	$2[1-T_{\nu+1}(z_{1/2})]$	–
Hüsler–Reiss	$2[1-\Phi(\frac{1}{\lambda})]$	–
Marshall–Olkin	$\min\{\alpha_1,\alpha_2\}$	–

should be ellipses with a center at $(0,0)$. For a visual inspection, we use contours of the function $g(z_1,z_2)$ to be studied, i.e., $g(z_1,z_2)=k$ for different values of k. In the above case, this provides an easy way to assess departures from the Gaussian copula assumption. The contour plots of a bivariate density obtained from a copula density transformed to achieve standard normal margins we call *normalized bivariate copula contour plots*. In general, this gives us three variables scales.

Definition 3.11 (*Variable scales*) We consider the following scales:

- **x-scale**: *original scale* (X_1,X_2) with density $f(x_1,x_2)$,
- **u-scale**: *copula scale* (U_1,U_2) where $U_i := F_i(X_i)$ and copula density $c(u_1,u_2)$, and
- **z-scale**: *marginal normalized scale* (Z_1,Z_2), where $Z_i := \Phi^{-1}(U_i) = \Phi^{-1}(F_i(X_i))$ for $i=1,2$ with density

$$g(z_1,z_2)=c(\Phi(z_1),\Phi(z_2))\phi(z_1)\phi(z_2).$$

Here $\Phi(\cdot)$ and $\phi(\cdot)$ are the distribution and density function of a $N(0,1)$ variable.

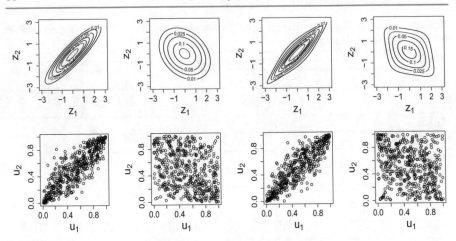

Fig. 3.8 `Bivariate elliptical copulas`: First column: Gauss with $\tau = .7$, second column: Gauss with $\tau = -.2$, third column: Student's t with $\nu = 4, \tau = .7$, and fourth column: Student's t with $\nu = 4, \tau = -.2$ (top row: normalized bivariate copula contours of $g(\cdot, \cdot)$ and bottom row: pairs plot of a random sample (u_{i1}, u_{i2}) on the copula scale)

Normalized contour plots for many parametric copula families are implemented in the R package `VineCopula` of Schepsmeier et al. (2018). For this, a `BiCop` object for the desired copula family and its parameter has to be generated and then the function `contour` can be applied.

In the following, we first give the normalized contour plots of the density $g(z_1, z_2)$ arising from bivariate elliptical copulas (Fig. 3.8). Here Gaussian copulas with $\tau = .7$ and $\tau = .2$ are compared to the corresponding Student's t copula with degree of freedom set to $\nu = 4$. A similar plot is given for the bivariate Clayton and Gumbel copulas in Fig. 3.9. These plots also show samples of the copula of size $n = 500$ in the bottom panels.

To explore the contour shapes for extreme-value copulas we also visualize Tawn copula for different combinations of the three parameters given in Fig. 3.10. The exploration of the normalized contours for further bivariate copula families is left as an exercise.

As expected the normalized bivariate copula contours in the two left panels of Fig. 3.8 are ellipses, while diamond shapes occur in the Student's t copula case as seen in two right panels of Fig. 3.8. A careful inspection of the sample scatter plots in Fig. 3.8 corresponding to the Student's t copula also shows that the joint occurrence of very large (close to 1) copula data values and very small (close to zero) values is larger when compared to the Gaussian copula case. The Clayton copula (Fig. 3.9) can be identified by lower tail dependence in the positive dependence case ($\tau > 0$), while the Gumbel copula (Fig. 3.9) shows the opposite behavior. For negative Kendall's τ values in the Clayton and Gumbel case we used a $90°$ rotation. The extreme-value Tawn copula allows for nonsymmetric shapes around the diagonal in contrast to the

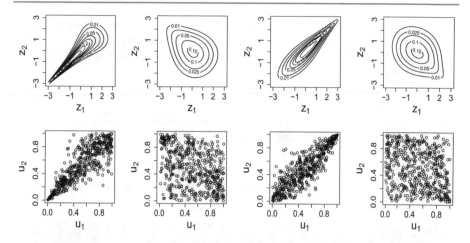

Fig. 3.9 Bivariate Archimedean copulas: First column: Clayton with $\tau = .7$, second column: Clayton with $\tau = -.2$, third column: Gumbel with $\tau = .7$, and fourth column: Gumbel with $\tau = -.2$ (top row: normalized bivariate copula contours of $g(\cdot, \cdot)$ and bottom row: pairs plot of a random sample (u_{i1}, u_{i2}) on the copula scale)

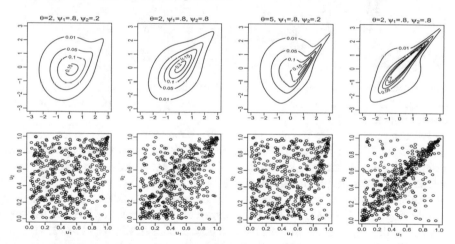

Fig. 3.10 Bivariate Tawn copula with three parameters: First column: $\tau = .14$, second column: $\tau = .38$, third column: $\tau = .18$, and fourth column: $\tau = .56$ (top row: normalized bivariate copula contours of $g(\cdot, \cdot)$ and bottom row: pairs plot of a random sample (u_{i1}, u_{i2}) on the copula scale)

other visualized copulas in Figs. 3.8 and 3.9. The resulting Kendall's τ values are also given in the caption of Fig. 3.10.

These contour shapes can now be utilized in an exploratory fashion to select appropriate copula families to observed copula data $\{(u_{i1}, u_{i2})^{\top}, i = 1, ...n\}$. In this case, we transform these observations to the z-scale by setting $(z_{i1}, z_{i2})^{\top} :=$ $(\Phi^{-1}(u_{i1}), \Phi^{-1}(u_{i2}))^{\top}$ for $i = 1, ...n$. Bivariate kernel density smoothing can now be applied to this data and used to estimate the corresponding contours. These *empirical normalized contours* can then be compared to the contour shapes arising from the parametric copula families. This allows for a first selection of an appropriate copula family to the data at hand.

3.9 Simulation of Bivariate Copula Data

Since the joint bivariate copula density $c_{12}(u_1, u_2)$ can be expressed as the product of $c_{1|2}(u_1|u_2)$ times $c_2(u_2) = 1$ for all (u_1, u_2), bivariate samples can be generated as follows.

> **Algorithm 3.12** (*Bivariate copula simulation*) To simulate i.i.d. data $\{(u_{i1}, u_{i2}),$
> $i = 1, ..., n\}$ from bivariate copula distributions
>
> - First, simulate u_{i1} from a uniform distribution on $[0, 1]$ for $i = 1, \ldots, n$.
> - Second, simulate u_{i2} from the conditional distribution $C_{2|1}(\cdot|u_{i1})$, where $C_{2|1}(\cdot|u_{i1}) = h_{2|1}(\cdot|u_{i1})$ as defined in Definition 1.16, i.e., set $u_{i2} = h_{2|1}^{-1}(v_{i2}|u_{i1})$ where v_{i2} is a realization of the uniform distribution for $i = 1, \ldots, n$.

This simulation algorithm is implemented by the function `BiCopSim` of the R package `VineCopula` of Schepsmeier et al. (2018) for many parametric bivariate copula families.

We can now compare theoretical normalized contour plots with ones based on samples. For this, we use *two-dimensional kernel density smoothing* as provided for example by the R function `kdecop` a of the R package `kdecopula` described by Nagler (2016). In particular, the local likelihood estimator with nearest neighbor bandwidth on the z-scale proposed by Geenens et al. (2017) is implemented in `kdecopula`. This method has been shown to be the preferred method for copula kernel density estimations of bivariate copula densities in Nagler (2014). This approach is also implemented in `VineCopula` for `copuladata` objects using the function `pairs`.

Remark 3.13 (*Bivariate kernel density estimates as estimates of the normalized z-scale density $g(\cdot, \cdot)$*) If we use copula data and transform it to the z-scale, then the corresponding sample $(z_{i1}, z_{i2}) = (\Phi^{-1}(u_{i1}), \Phi^{-1}(u_{i1}))$ is a sample which has standard normal margins. However, general kernel density estimation methods such as `kde2d` do not enforce this restriction on the margins; therefore, they are only approximations to the density corresponding to the z-data.

We will now illustrate how the empirical normalized contour plot based on bivariate kernel density estimation compares to theoretical normalized contour plot for

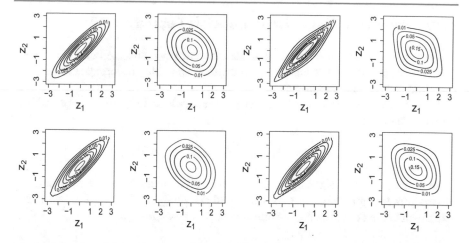

Fig. 3.11 Bivariate elliptical copulas: First column: Gauss with $\tau = .7$, second column: Gauss with $\tau = -.2$, third column: Student's t with $\nu = 4, \tau = .7$, and fourth column: Student's t with $\nu = 4, \tau = -.2$ (top row: normalized bivariate copula contours of $g(\cdot, \cdot)$ and bottom row: empirical normalized bivariate copula contours based on a sample of size $n = 500$)

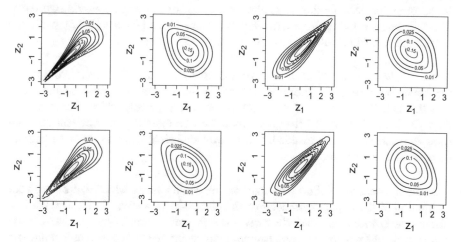

Fig. 3.12 Bivariate Archimedean copulas: First column: Clayton with $\tau = .7$, second column: Clayton with $\tau = -.2$, third column: Gumbel with $\tau = .7$, and fourth column: Gumbel with $\tau = -.2$ (top row: normalized bivariate copula contours of $g(\cdot, \cdot)$ and bottom row: empirical normalized bivariate copula contours based on a sample of size $n = 500$)

known parametric bivariate copula families. In Fig. 3.11 we compare empirical normalized contour plots for medium and high dependence as measured by Kendall's τ using `BiCopKDE` in the R package `VineCopula` with theoretical contour plots for two elliptical bivariate copula families. For the bivariate Archimedean Clayton and Gumbel copula the corresponding plot is given in Fig. 3.12. Both plots illustrate the satisfactory performance of the kernel density smoothing for this sample size.

3.10   Parameter Estimation in Bivariate Copula Models

Assume that you have bivariate independent identically distributed data $\{(x_{i1}, x_{i2}),$ $i = 1, ..., n\}$ available. The marginal distributions can either be known, known up to marginal parameters or unknown.

In the known case we transform directly to the copula scale by using the probability integral transform

$$(u_{i1}, u_{i2}) := (F_1(x_{i1}), F_2(x_{i2})) \text{ for } i = 1, ..., n.$$

In the case of unknown margins, one often uses a *two-step approach* by estimating the margins first and then using the estimated marginal distributions $\hat{F}_j$, $j = 1, 2$ to transform to the copula scale by defining the *pseudo-copula data*

$$(u_{i1}, u_{i2}) := (\hat{F}_1(x_{i1}), \hat{F}_2(x_{i2})) \text{ for } i = 1, ..., n.$$

and then formulate a copula model for the pseudo-copula data. If parametric marginal models are used, then we speak of an *inference for margins approach* (IFM) and if the empirical distribution is applied we speak of a *semiparametric approach*. The efficiency of the IFM approach has been investigated by Joe (2005), while the semiparametric approach has been proposed by Genest et al. (1995).

Remark 3.14 (Misspecifications of marginal distributions) Kim et al. (2007) investigated in simulations the error one makes if the margins are misspecified and found that this is only a problem if the margins are severely misspecified. For example, they studied fitting normal margins, when the true margins are exponential. Such misspecifications can be avoided, if a careful statistical analysis for the marginal distributions is performed.

In both cases, we end up with data on the copula scale, which can be used for estimation of the copula parameter of the chosen bivariate copula family. We first discuss the *inversion of Kendall's τ approach* for estimation. In this case, we choose a copula family which has a one-to-one relationship between the copula parameter θ and Kendall's τ, i.e.,

$$\tau = k(\theta)$$

we can use the empirical estimator $\hat{\tau}$ of τ defined in (2.4) or (2.5) and set

$$\hat{\theta}_\tau := k^{-1}(\hat{\tau}).$$

This gives an estimator of the copula parameter θ. Its asymptotic distribution you can determine using the Delta method applied to the asymptotic distribution of $\hat{\tau}$ as given in Theorem 2.8. This is valid as long as the marginal distributions are known up to the parameter and only approximately valid in the case of unknown margins.

Table 3.2 gives examples of bivariate parametric copula families that are amenable to this way of proceeding. In the `VineCopula` package, the R function `BiCopEst` with option `method="itau"` uses this inversion approach.

Alternatively, we can use maximum likelihood (ML), since the likelihood is given

$$\ell(\theta; \boldsymbol{u}) = \prod_{i=1}^{n} c(u_{i1}, u_{i2}; \theta), \tag{3.21}$$

where $\boldsymbol{u} := \{(u_{i1}, u_{i2}), i = 1, ..., n\}$ is the observed or pseudo-copula data. We speak of `pseudo-maximum likelihood` if pseudo-copula data is used. The (pseudo) maximum likelihood estimator $\hat{\theta}_{ML}$ maximizes (3.21). This approach is implemented in `VineCopula` by using `BiCopEst` with `method="mle"`.

This covers the case of independent identically distributed data $\{(x_{i1}, x_{i2}), i = 1, ..., n\}$. However, it there are marginal time series structures and/or regression structures in the marginal data, then these have to be accounted for using appropriate marginal models.

Remark 3.15 (*Handling marginal data structures*) In the case of marginal time series structures ARMA(p,q) and/or GARCH(p,q) models (see, for example, Ruppert 2004) can be fitted first to each of the two margins to account for marginal time dependencies, and in a second step, standardized residuals are formed. These are then approximately independent identically distributed and thus can be used to transform to copula data. For the inclusion of covariate effects appropriate regression models can be used and again the resulting standardized residuals are used for further processing.

Ex 3.4 (`WINE3`: *Empirical and fitted normalized contour plots*) The original data is transformed to copula data scale using marginal empirical distributions.

Now we conduct a pairwise analysis to explore the dependence structure among all pairs of variables. First, we determined empirical contours to the normalized z-data using the function `pairs` in the R library `VineCopula` of Schepsmeier et al. (2018), where the data is stored in a `copuladata` object. The results are displayed in Fig. 3.13 and show non-elliptical dependencies. Comparing to the shapes given in Sect. 3.8 a rotated 270° Gumbel copula might be appropriate for the pair (`acf`, `acv`), while a Gumbel copula for the pair (`acf`, `acc`) and a Frank copula for the pair (`acv`, `acc`) might be reasonable choices, respectively. The shape of the Frank copula will be explored in Exercise 3.7. The estimated Kendall's τ values, the selected copula family and the (through

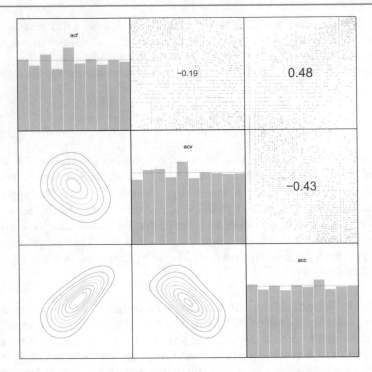

Fig. 3.13 WINE3: Upper triangle: pairs plots of copula data, diagonal: Marginal histograms of copula data, and lower triangle: empirical contour plots of normalized copula data

Table 3.4 WINE3 : Estimated Kendall's τ, chosen copula family, and through inversion estimated copula parameter for all pairs of variables

Pair	$\hat{\tau}$	Copula family	$\hat{\theta}$
(acf, acv)	−0.19	Gumbel 270	−1.2
(acf, acc)	0.48	Gumbel	1.9
(acv, acc)	−0.43	Frank	−4.6

inversion) estimated copula parameter are given in Table 3.4. To check informally our choices we use the estimated Kendall's τ values given in Table 2.1 and transform them to a corresponding parameter estimate using the relationships given in Table 3.2. The corresponding fitted normalized contour plots are given in Fig. 3.14.

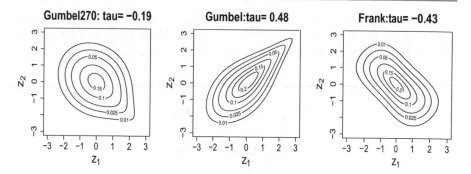

Fig. 3.14 WINE3: Fitted normalized contour plots for the chosen bivariate copula family with parameter determined by the empirical Kendall's τ estimate (left: acf versus acv, middle: acf versus acc, and right: acv versus acc)

3.11 Conditional Bivariate Copulas

Sometimes one is interested in the dependence of two variables (X_1, X_2) conditional on the value of a third variable $X_3 = x_3$. This dependence might change as the value of x_3 is changing. This dependence is characterized by the copula corresponding to the bivariate distribution of (X_1, X_2) conditional on $X_3 = x_3$.

We start our discussion with the special case where three variables (U_1, U_2, U_3) have uniform margins and their joint distribution is given by a three-dimensional copula C_{123} with copula density

$$c_{123}(u_1, u_2, u_3) = \frac{\partial^3}{\partial u_1 \partial u_2 \partial u_3} C_{123}(u_1, u_2, u_3).$$

To determine the bivariate distribution function $C_{12|3}(\cdot, \cdot | v_3)$ of (U_1, U_2) given $U_3 = v_3$ we integrate the conditional density $c_{12|3}(u_1, u_2 | v_3) = c_{123}(u_1, u_2, v_3)$ to get

$$\begin{aligned}
C_{12|3}(u_1, u_2 | v_3) &= \int_0^{u_1} \int_0^{u_2} c_{123}(v_1, v_2, v_3) dv_1 dv_2 \\
&= \int_0^{u_1} \int_0^{u_2} \frac{\partial^3}{\partial v_1 \partial v_2 \partial v_3} C_{123}(v_1, v_2, v_3) dv_1 dv_2 \\
&= \frac{\partial}{\partial u_3} C_{123}(u_1, u_2, u_3)|_{u_3 = v_3}.
\end{aligned} \tag{3.22}$$

To derive the copula $C_{12;3}$ corresponding to the bivariate distribution (U_1, U_2) given $U_3 = v_3$ we will apply the inverse statement of Sklar's Theorem (1.17) to this distribution. Thus we require the marginal distribution functions of (U_1, U_2) given $U_3 = v_3$. We denote by $C_{i(3)}(c_{i(3)})$ the distribution (density) of the ith margin of (U_1, U_2) given $U_3 = v_3$ for $i = 1, 2$. In particular, we have using (3.22)

$$C_{1(3)}(u_1|v_3) = C_{12|3}(u_1, 1|u_3)$$
$$= \frac{\partial}{\partial u_3} C_{123}(u_1, 1, u_3)|_{u_3=v_3}$$
$$= \frac{\partial}{\partial u_3} C_{13}(u_1, u_3)|_{u_3=v_3} = C_{1|3}(u_1|v_3).$$

Similarly, we get $C_{2(3)}(u_2|v_3) = C_{2|3}(u_2|v_3)$ showing that the margins of (U_1, U_2) given $U_3 = v_3$ agree with the conditional distribution functions of U_i given $U_3 = v_3$ for $i = 1, 2$. Now we have calculated all building blocks for the inverse statement of Sklar's theorem (1.17) to derive the copula $C_{12;3}$ of (U_1, U_2) given $U_3 = v_3$ with distribution function $C_{12|3}$, which is given by

$$C_{12;3}(u_1, u_2|v_3) = C_{12|3}(C_{1|3}^{-1}(u_1|v_3), C_{2|3}^{-1}(u_2|v_3)|v_3). \qquad (3.23)$$

We speak then of the *conditional bivariate copula*.

For a general bivariate distribution X_1, X_2 conditional on $X_3 = x_3$ we have the same copula specified in (3.23) with conditioning value $v_3 = F_3(x_3)$, where F_3 is the marginal distribution function of X_3. This follows from the invariance property of copulas given in Lemma 1.13.

Ex 3.5 (Conditional copula associated with a trivariate Frank copula) The *trivariate* extension of the *Frank* copula considered in (3.5) is given

$$C_{123}(u_1, u_2, u_3; \delta) = \frac{1}{\ln(\alpha)} \ln\left[1 - (1 - \alpha) \prod_{i=1}^{3} \frac{1 - \alpha^{u_i}}{1 - \alpha}\right], \qquad (3.24)$$

where $\alpha := \exp(-\delta)$ for $\delta \in [-\infty, \infty]\setminus\{0\}$. We now want to determine the bivariate distribution function of (U_1, U_3) given $U_2 = v_2$. Straightforward differentiation and applying (3.22) determines this conditional distribution as

$$C_{13|2}(u_1, u_3|v_2; \delta) = \frac{\partial}{\partial u_2} C_{123}(u_1, u_2, u_3; \theta)|_{u_2=v_2}$$
$$= \frac{-\alpha^{v_2}(1 - \alpha^{u_1})(1 - \alpha^{u_3})}{(1 - \alpha)^2 - \prod_{i=1}^{3}(1 - \alpha^{u_i})|_{u_2=v_2}}. \qquad (3.25)$$

To determine the associated conditional copula $C_{13;2}(u_1, u_3 | u_2 = v_2; \delta)$ to (3.25) we need the following marginals:

$$C_{1|2}(u_1|v_2; \delta) = \frac{\partial}{\partial u_2} C_{12}(u_1, u_2; \delta)|_{u_2=v_2} = -\frac{\alpha^{v_2}(1 - \alpha^{u_1})}{(1 - \alpha) - (1 - \alpha^{u_1})(1 - \alpha^{v_2})},$$

$$C_{3|2}(u_3|v_2; \delta) = \frac{\partial}{\partial u_2} C_{32}(u_3, u_2; \delta)|_{u_2=v_2} = -\frac{\alpha^{v_2}(1 - \alpha^{u_3})}{(1 - \alpha) - (1 - \alpha^{u_3})(1 - \alpha^{v_2})}.$$

To apply (3.23), we need to invert the above margins by setting $p_i = C_{i|2}(u_i|u_2; \delta) \in (0, 1)$ for $i = 1, 3$. In particular, this implies

$$(1 - \alpha^{u_i}) = -\frac{p_i(1 - \alpha)}{\alpha^{v_2} - p_i(1 - \alpha^{v_2})}$$

for $i = 1, 3$. Using these expressions, it follows that the conditional copula $C_{13;2}(p_1, p_3|v_2; \delta)$ can be expressed as

$$C_{13;2}(p_1, p_3|v_2; \delta) = C_{13|2}(C_{1|2}^{-1}(p_1|v_2; \delta), C_{3|2}^{-1}(p_3|v_2; \delta)|v_2; \delta)$$

$$= \frac{p_1 p_3}{(1 - \alpha^{v_2})[1 - (1 - p_1)(1 - p_3)] - \alpha^{v_2}}. \quad (3.26)$$

Comparing (3.25) with (3.26), we see that the bivariate distribution of (U_1, U_3) given $U_2 = v_2$ does not agree with the associated conditional copula. Thus, we use different notations such as $C_{13|2}$ versus $C_{13;2}$.

Ex 3.6 (Conditional copula associated with a trivariate Clayton copula) The *trivariate* extension of the *Clayton copula* considered in (3.3) is given by

$$C_{123}(u_1, u_2, u_3; \delta) = (u_1^{-\delta} + u_2^{-\delta} + u_3^{-\delta} - 2)^{-\frac{1}{\delta}}, \quad (3.27)$$

for $0 < \delta < \infty$. For (U_1, U_2, U_3) with distribution function (3.27), the distribution function of (U_1, U_3) given $U_2 = v_2$ is given by

$$C_{13|2}(u_1, u_3|v_2; \delta) = (u_1^{-\delta} + v_2^{-\delta} + u_3^{-\delta} - 2)^{-\frac{1+\delta}{\delta}} v_2^{-(\delta+1)}, \quad (3.28)$$

with corresponding margins

$$C_{1|2}(u_1|v_2; \delta) = (1 + \delta)(u_1^{-\delta} + v_2^{-\delta} - 1)^{-\frac{\delta+1}{\delta}} v_2^{-(\delta+1)} \quad (3.29)$$

and similar expression for $C_{1|3}$.

Proceeding as in Example 3.5, we can show that the copula associated with (U_1, U_3) given $U_2 = v_2$ is given by

$$C_{13;2}(p_1, p_3 | v_2; \delta) = \left(p_1^{-\frac{\delta}{1+\delta}} + p_3^{-\frac{\delta}{1+\delta}} - 1 \right)^{-\frac{\delta+1}{\delta}}. \qquad (3.30)$$

This shows that the conditional copula is independent of the conditioning value v_2 and can be identified as a bivariate Clayton copula with parameter $\frac{\delta}{1+\delta}$.

3.12 Average Conditional and Partial Bivariate Copulas

In general, the conditional copula $C_{13;2}$ as defined in (3.23) will depend on the conditioning value, and thus, Gijbels et al. (2015) considered the *average conditional copula* defined as

$$C_{13;2}^A(u_1, u_3) := \int_0^1 C_{13;2}(u_1, u_3 | v_2) dv_2, \qquad (3.31)$$

where the dependency on the conditioning value is integrated out. This concept was first mentioned in Bergsma (2004, 2011) in the context of nonparametric tests for conditional independence. From this definition, it is not clear that the average conditional copula is a copula. We will show this later.

Another approach to remove the dependency on the conditioning value is the *partial copula*, which was also discussed in Bergsma (2004). For this, we consider the random variables

$$V_{1|2} := C_{1|2}(U_1 | U_2) \text{ and } V_{3|2} := C_{3|2}(U_3 | U_2). \qquad (3.32)$$

These random variables are also called *conditional probability integral transforms (CPIT)*. The distribution of $V_{1|2}$ is uniform since

$$\begin{aligned}
P(V_{1|2} \le v_{1|2}) &= \int_0^1 P(C_{1|2}(U_1 | U_2) \le v_{1|2} | U_2 = u_2) du_2 \\
&= \int_0^1 P(U_1 \le C_{1|2}^{-1}(v_{1|2} | u_2) | U_2 = u_2) du_2 \\
&= \int_0^1 C_{1|2}(C_{1|2}^{-1}(v_{1|2} | u_2) | u_2) du_2 \\
&= \int_0^1 v_{1|2} du_2 = v_{1|2}
\end{aligned}$$

holds. Similarly, $V_{3|2}$ has a uniform distribution. Therefore, the joint distribution of $(V_{1|2}, V_{3|2})$ is a copula and this copula is called the *partial copula* and is denoted by C^P. Note that this copula does not depend on a specific value the conditioning variable U_2.

Following Gijbels et al. (2015) we show now that the partial and the average conditional copula coincides. For this, consider

$$C^P(v_{1|2}, v_{3|2}) := P(V_{1|2} \leq v_{1|2}, V_{3|2} \leq v_{3|2})$$

$$= \int_0^1 P(C_{1|2}(U_1|U_2) \leq v_{1|2}, C_{3|2}(U_3|U_2) \leq v_{3|2}|U_2 = u_2)du_2$$

$$= \int_0^1 P(U_1 \leq C_{1|2}^{-1}(v_{1|2}|u_2), U_3 \leq C_{3|2}^{-1}(v_{3|2}|u_2)|U_2 = u_2)du_2$$

$$= \int_0^1 C_{13|2}(C_{1|2}^{-1}(v_{1|2}|u_2), C_{3|2}^{-1}(v_{3|2}|u_2)|U_2 = u_2)du_2$$

$$= \int_0^1 C_{13;2}(C_{1|2}(C_{1|2}^{-1}(v_{1|2}|u_2)|u_2), C_{3|2}(C_{3|2}^{-1}(v_{3|2}|u_2)|u_2)|u_2)du_2$$

$$= \int_0^1 C_{13;2}(v_{1|2}, v_{3|2}|u_2)du_2$$

$$= C^A(v_{1|2}, v_{3|2})$$

for all values $v_{1|2}$ and $v_{3|2}$ in $(0, 1)$. This also shows that the average conditional copula is a copula, since the partial copula is a copula by definition.

3.13 Exercises

Exer 3.1
Galambos copula: Derive from the Pickands dependence function given in Table 3.1 the Galambos copula and its copula density. Visualize this copula using normalized contour plots for several values of δ.

Exer 3.2
t-EV copula: Derive from the Pickands dependence function given in Table 3.1 the t-EV copula and its copula density. Visualize this copula using normalized contour plots for $\rho = .3$ and $\rho = .7$ for $\nu = 5$.

Exer 3.3
Exploratory bivariate copula choices for the seven-dimensional red wine data: For the data set considered in Exercise 1.7 the pairs plot of the associated pseudo-copula data is given in Fig. 3.15. For each pair of variables propose a pair copula family.

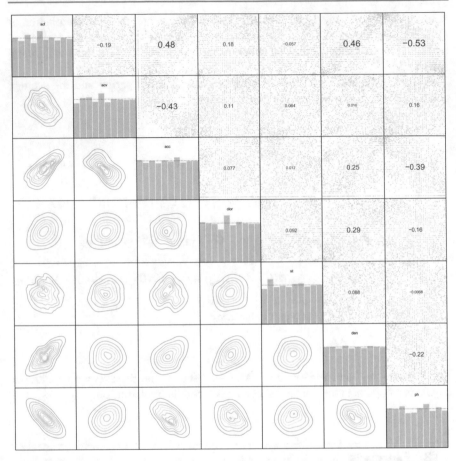

Fig. 3.15 WINE7: upper triangular: pairs plots of copula data, diagonal: Marginal histograms of copula data, lower triangular: empirical contour plots of normalized copula data

Exer 3.4

URAN3: *Exploratory copula choices for the three-dimensional* uranium *data*: Consider as in Example 2.2 the three-dimensional subset of the uranium data set contained in the R package copula with variables Cobalt (Co), Titanium (Ti) and Scandium (Sc). As in Example 3.4 transform the original data to the copula scale using marginal empirical distributions. Then explore the empirical normalized contour plots for all pairs of variables and suggest appropriate parametric pair copula families. Check your choices by comparing the fitted to the empirical normalized contour plots.

Exer 3.5

ABALONE3: *Exploratory copula choices for the three-dimensional* abalone *data*: Consider as in Example 2.3 the three-dimensional subset of the abalone data set contained in the R package `PivotalR` with variables `shucked`, `viscera`, and `shell`. As in Example 3.4 transform the original data to the copula scale using marginal empirical distributions. Then explore the empirical normalized contour plots for all pairs of variables and suggest appropriate parametric pair copula families. Check your choices by comparing the fitted to the empirical normalized contour plots.

Exer 3.6

The effect of the degree of freedom in a bivariate Student's t copula on the contour shapes: For $df = 2, ..., 30$ draw the normalized contour plots, when the association parameter is $\rho = .7$. Do the same for $\rho = -.2$. How do these plots change when you fix $\tau = .7$ and $\tau = -.2$, respectively.

Exer 3.7

Normalized contour shapes for Frank and Joe copulas: Draw normalized contour shapes for the bivariate Frank and Joe copula defined in Example 3.1 for Kendall's τ values .7 and $-.2$ using the function `contour` in the R package `VineCopula`. Note you might need to apply rotations.

Exer 3.8

Normalized contour shapes for two parameter subclasses of the Tawn copula with Pickands dependence function given in Table 3.1: Draw normalized contour shapes for the Tawn copula with two parameters. In particular, in the R package `VineCopula`, the family `fam=104` represents the Tawn copula where $\psi_2 = 1$, while `fam=204` sets $\psi_1 = 1$. Again use `contour` in the R package `VineCopula`. Explore also the range of the Kendall's τ obtained as the parameters of these two parameter subfamilies vary. Do the same for the upper tail dependence coefficient.

Exer 3.9

Kendall's τ values and normalized contour shapes for BB1 and BB7 copulas:

- Draw perspective plots for Kendall's τ values as the two parameters (θ, δ) of the BB1 family change. Do the same for the BB7 family.
- Draw perspective plots for the upper tail dependence coefficient values as the two parameters (θ, δ) of the BB1 family change. What is the range of achievable upper tail dependence coefficients? Do the same for the BB7 family.
- Draw perspective plots for the lower tail dependence coefficient values as the two parameters (θ, δ) of the BB1 family change. What is the range of achievable lower tail dependence coefficients? Do the same for the BB7 family.
- Draw normalized contour shapes for the bivariate BB1 and BB7 copula defined in Example 3.2 for two parameter configurations of θ and δ yielding a Kendall's τ value .7. Do the same for a resulting Kendall's τ value of -0.2. Again rotations might be necessary.

Exer 3.10

Conditional distribution of the Clayton copula: Derive and visualize the h functions $C_{2|1}(u_2|u_1 = .5)$ and $C_{1|2}(u_1|u_2 = .5)$ of a bivariate Clayton copula with a Kendall's $\tau = .5$ and $\tau = .8$, respectively. Compare the two functions. Do the same for a 90° rotated Clayton copula.

Exer 3.11

Conditional distribution of the Gumbel copula: Derive and visualize graphically the h functions $C_{2|1}(u_2|u_1 = .75)$ and $C_{1|2}(u_1|u_2 = .75)$ of a bivariate Gumbel copula with a Kendall's $\tau = .7$ and $\tau = .2$, respectively. Compare the two functions. Do the same for a 270° rotated Gumbel copula.

Exer 3.12

Conditional distribution of the Frank copula: Derive and visualize graphically the h functions $C_{2|1}(u_2|u_1 = .25)$ and $C_{1|2}(u_1|u_2 = .25)$ of a bivariate Frank copula for with a Kendall's $\tau = .7$ and $\tau = .2$, respectively. Compare the two functions. Do the same for a 90° rotated Frank copula.

Exer 3.13

Bivariate survival copula: Prove the relationship (3.20) for bivariate survival times.

Exer 3.14

Rotated and reflected versions of the two parameter Tawn copula:

- Consider again the two parameter Tawn copula family of Exercise 3.8. For this family plot normalized contour plots of all rotations and all reflections discussed in Sect. 3.6.

- Under which conditions on the bivariate copula density can 1-reflected copula density be interpreted as a rotated copula density?

- Show that $C^{*1}(u_1, u_2) = u_2 - C(1 - U_1, u_2)$ is a copula and that it is the copula of $(1 - U_1, U_2)$.

Exer 3.15

Conditional copula associated with a trivariate Clayton copula: For Example 3.6 derive Eqs. (3.28), (3.29) and (3.30).

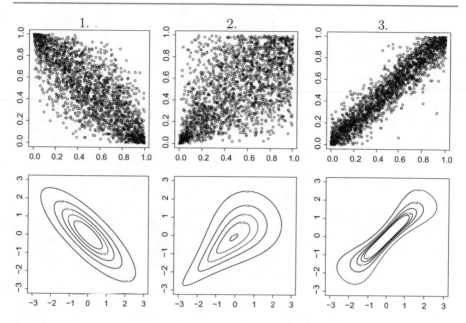

Fig. 3.16 Exploratory visualization: Top row: scatter plots on u-level scale. Bottom row: normalized contour plots on z-level scale

Exer 3.16

Bivariate scatter and normalized contour plots: Analyze the bivariate scatter plots of u-level and corresponding normalized contour plots on the z-level given in Fig. 3.16 with respect to the strength of dependence, tail dependencies, symmetries, and exchangeability. In addition, for each of the three panels name a possible parametric pair copula family that can be used to model the described dependence characteristics.

Pair Copula Decompositions and Constructions

4

The goal is to construct multivariate distributions using only bivariate building blocks. The appropriate tool to obtain such a construction is to use conditioning. Joe (1996) gave the first *pair copula construction* of a multivariate copula in terms of distribution functions, while Bedford and Cooke (2001, 2002) independently developed constructions expressed in terms of densities. Additionally they provided a general framework to identify all possible constructions.

4.1 Illustration in Three Dimensions

We start with random variables X_1, X_2 and X_3 and consider the recursive factorization of their joint density given by

$$f(x_1, x_2, x_3) = f_{3|12}(x_3|x_1, x_2) f_{2|1}(x_2|x_1) f_1(x_1). \qquad (4.1)$$

Now we consider each part separately. To determine $f_{3|12}(x_3|x_1, x_2)$ we consider the bivariate conditional density $f_{13|2}(x_1, x_3|x_2)$. Note that this density has $F_{1|2}(x_1|x_2) (f_{1|2}(x_1|x_2))$ and $F_{3|2}(x_3|x_2) (f_{3|2}(x_3|x_2))$ as marginal distributions (densities) with associated copula density $c_{13;2}(\cdot, \cdot; x_2)$. More specifically, $c_{13;2}(\cdot, \cdot; x_2)$ denotes the copula density associated with the conditional distribution of (X_1, X_3) given $X_2 = x_2$. By Sklar's Theorem 1.9 we have

$$f_{13|2}(x_1, x_3|x_2) = c_{13;2}(F_{1|2}(x_1|x_2), F_{3|2}(x_3|x_2); x_2) f_{1|2}(x_1|x_2) f_{3|2}(x_3|x_2). \qquad (4.2)$$

© Springer Nature Switzerland AG 2019
C. Czado, *Analyzing Dependent Data with Vine Copulas*, Lecture Notes in Statistics 222, https://doi.org/10.1007/978-3-030-13785-4_4

Now $f_{3|12}(x_3|x_1, x_2)$ is the conditional density of X_3 given $X_1 = x_1$, $X_2 = x_2$ which can be determined using Lemma 1.15 applied to (4.2) yielding

$$f_{3|12}(x_3|x_1, x_2) = c_{13;2}(F_{1|2}(x_1|x_2), F_{3|2}(x_3|x_2); x_2) f_{3|2}(x_3|x_2). \qquad (4.3)$$

Finally direct application of Lemma 1.15 gives

$$f_{2|1}(x_2|x_1) = c_{12}(F_1(x_1), F_2(x_2)) f_2(x_2) \qquad (4.4)$$
$$f_{3|2}(x_3|x_2) = c_{23}(F_2(x_2), F_3(x_3)) f_3(x_3) \qquad (4.5)$$

Inserting (4.3), (4.4) and (4.5) into (4.1) yields a pair copula decomposition of the joint density.

> **Definition 4.1** (*A pair copula decomposition in three dimensions*) A *pair copula decomposition* of an arbitrary three dimensional density is given as
>
> $$f(x_1, x_2, x_3) = c_{13;2}(F_{1|2}(x_1|x_2), F_{3|2}(x_3|x_2); x_2) \times c_{23}(F_2(x_2), F_3(x_3)) \quad (4.6)$$
> $$\times c_{12}(F_1(x_1), F_2(x_2)) f_3(x_3) f_2(x_2) f_1(x_1).$$

From this decomposition we see that the joint three dimensional density can be expressed in terms of bivariate copulas and conditional distribution functions. However this decomposition is not unique, since

$$f(x_1, x_2, x_3) = c_{12;3}(F_{1|3}(x_1|x_3), F_{2|1}(x_2|x_1); x_3) \times c_{13}(F_1(x_1), F_3(x_3))$$
$$\times c_{23}(F_2(x_2), F_3(x_3)) f_3(x_3) f_2(x_2) f_1(x_1) \qquad (4.7)$$

and

$$f(x_1, x_2, x_3) = c_{23;1}(F_{2|1}(x_2|x_1), F_{3|1}(x_3|x_1); x_1) \times c_{13}(F_1(x_1), F_3(x_3)) \quad (4.8)$$
$$\times c_{12}(F_1(x_1), F_2(x_2)) f_3(x_3) f_2(x_2) f_1(x_1)$$

are two different decomposition, which result from a reordering of the variables in (4.1).

We speak of a *pair copula decomposition* of a multivariate distribution, when the copulas associated with conditional distributions are allowed to depend on the specific value of the underlying conditioning variable. In (4.6) this means that $c_{13;2}(\cdot, \cdot; x_2)$ depends on x_2. In the following we make often the assumption that this dependence can be ignored. In this case we speak of making the *simplifying assumption*. For a in-depth discussion of the simplifying assumption see Sect. 5.4. In the remainder of this section we assume that the simplifying assumption holds. Mathematically the simplifying assumption in three dimensions is given as follows.

Definition 4.2 (*Simplifying assumption in three dimensions*) The simplifying assumption of a three dimensional pair copula construction based on (4.6) is satisfied when for any $x_2 \in \mathbb{R}$

$$c_{13;2}(u_1, u_2; x_2) = c_{13;2}(u_1, u_2) \text{ for } u_1 \in [0, 1], u_2 \in [0, 1]$$

holds.

Assuming the simplifying assumption, the right hand side of (4.6) can also be used in a constructive fashion: Specify arbitrary parametric bivariate copula families for $c_{12;3}$, c_{23} and c_{12} copula densities with parameters $\theta_{12;3}$, θ_{23} and θ_{12} respectively and define a valid parametric joint density as follows:

Definition 4.3 (*Pair copula construction of a joint parametric density in three dimensions*) A *parametric pair copula construction in three dimensions* specifies a three dimensional density with parameter vector $\theta = (\theta_{12}, \theta_{23}, \theta_{12;3})$ as

$$f(x_1, x_2, x_3; \theta) := c_{13;2}(F_{1|2}(x_1|x_2), F_{3|2}(x_3|x_2); \theta_{13;2}) \times c_{23}(F_2(x_2), F_3(x_3); \theta_{23})$$

$$\times c_{12}(F_1(x_1), F_2(x_2), \theta_{12}) f_3(x_3) f_2(x_2) f_1(x_1),$$

(4.9)

where $c_{13;2}(\cdot, \cdot; \theta_{13;2})$, $c_{12}(\cdot, \cdot; \theta_{12})$ and $c_{23}(\cdot, \cdot; \theta_{23})$ are arbitrary parametric bivariate copula densities.

In Definition 4.3 we also could allow for marginal parameters for the marginal densities f_1, f_2 and f_3. In the future we are often interested in considering the dependence structure as characterized by the copula on its own.

Definition 4.4 (*Pair copula construction of a three dimensional parametric copula*) A three dimensional parametric copula family with parameter vector $\theta = (\theta_{12}, \theta_{23}, \theta_{12;3})$ can be defined as follows

$$c(u_1, u_2, u_3; \theta) := c_{13;2}(C_{1|2}(u_1|u_2), C_{3|2}(u_3|u_2); \theta_{13;2}) \times c_{23}(u_2, u_3; \theta_{23})$$

$$\times c_{12}(u_1, u_2, \theta_{12}),$$

(4.10)

where $C_{1|2}(\cdot|u_2)$ and $C_{3|2}(\cdot|u_2)$ are the conditional distribution functions of U_1 given $U_2 = u_2$ and U_3 given $U_2 = u_2$, respectively.

Recall that the conditional distribution functions $C_{1|2}(\cdot|u_2)$ and $C_{3|2}(\cdot|u_2)$ needed in the construction have already been determined in in Eq. (1.26).

From the construction (4.9) or (4.10) we see that the bivariate marginal distribution of the variables (X_1, X_3) or (U_1, U_3) is not directly given, it can however be determined by integration. In particular we have for (4.10)

$$
c_{13}(u_1, u_3; \boldsymbol{\theta}) = \int_0^1 c_{13;2}(C_{1|2}(u_1|u_2; \theta_{12}), C_{3|2}(u_3|u_2; \theta_{23}); \theta_{13;2})
$$
$$
\times\, c_{23}(u_2, u_3; \theta_{23})c_{12}(u_1, u_2; \theta_{12})du_2.
$$

In practice we can also use simulated values to determine this bivariate margin. To explore the flexibility for the shape of this margin we studied in Chap. 5 of Stöber and Czado (2017) the parametric pair copula constructions (4.10) as specified in Table 4.1.

The panels in Fig. 4.1 demonstrate the flexibility of the pair copula constructions in 3 dimensions. They are based on two-dimensional kernel density estimates using 1000000 data points simulated from (4.10) transformed to standard normal margins. None of these normalized contour plots resemble the shapes of normalized contour plots from known parametric bivariate copulas discussed in Chap. 3.

In three dimensions we can also visualize three dimension contour regions attached to a three dimensional density for different levels k, where $f(x_1, x_2, x_3) = k$ holds. The contour regions on the normalized z-scale ($z_1 = \Phi^{-1}(u_1), z_2 = \Phi^{-1}(u_2), z_3 = \Phi^{-1}(u_3)$) were recently studied in Killiches et al. (2017) for arbitrary three dimensional PCC's. In Fig. 4.2 we show now the 3D normalized contour plots for three different angles (given in columns) corresponding to the cases specified in Table 4.1 (given in rows).

The non Gaussian behavior of the three variate copulas specified in Table 4.1 is clearly visible by the non ellipsoid shapes in Fig. 4.2.

Before we continue with the pair copula construction we would like to give further characterizations of the copula associated with the conditional distribution (X_1, X_3) given $X_2 = x_2$.

Table 4.1 Specifications: Models for the panels of normalized contour plots in Fig. 4.1 (Taken from Stöber and Czado (2017). With permission of ©World Scientific Publishing Co. Pte. Ltd. 2017.)

Case	(1, 2)			(2, 3)			(1, 3; 2)		
	Family	Par.	τ	Family	Par.	τ	Family	Par.	τ
1	Gumbel	5	0.80	Clayton	−0.7	−0.26	Clayton	0.7	0.26
2	Student t	(0.8, 1.2)	0.59	Gumbel	1.75	0.43	Student t	(−0.95, 2.5)	−0.80
3	Joe	7	0.76	Joe	24	0.92	Joe	4	0.61
4	Frank	−40	−0.90	Clayton	20	0.91	Frank	100	0.96

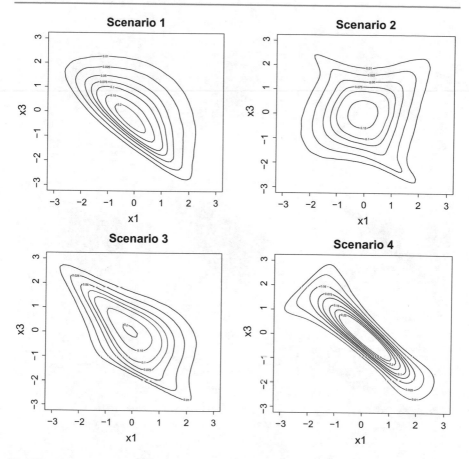

Fig. 4.1 Normalized contours: Estimated normalized contours of the bivariate $(1, 3)$ margin corresponding to the four cases given in Table 4.1 (Taken from Stöber and Czado (2017). With permission of ©World Scientific Publishing Co. Pte. Ltd. 2017.)

Ex 4.1 (Copula and density corresponding to (X_1, X_3) given $X_2 = x_2$) Assuming the simplifying assumption and using Sklar's Theorem 1.9 for the conditional distribution function (X_1, X_3) given X_2 denoted by $F_{13|2}$ we can express $C_{13;2}(\cdot, \cdot | x_2)$ the copula associated with the bivariate conditional distribution function $F_{13|2}$ as

$$C_{13;2}(u_1, u_3; x_2) = F_{13|2}(F_{1|2}^{-1}(u_1|x_2), F_{3|2}^{-1}(u_3|x_2)).$$

Therefore the corresponding density $c_{13;2}(\cdot, \cdot; x_2)$ satisfies

$$c_{13;2}(u_1, u_3|x_2) = \frac{\partial^2}{\partial v_1 \partial v_3} F_{13|2}(v_1, v_3|x_2)\big|_{v_1 = F_{1|2}^{-1}(u_1|x_2), v_3 = F_{3|2}^{-1}(u_3|x_2)} \frac{dv_1}{du_1}\frac{dv_3}{du_3} \quad (4.11)$$

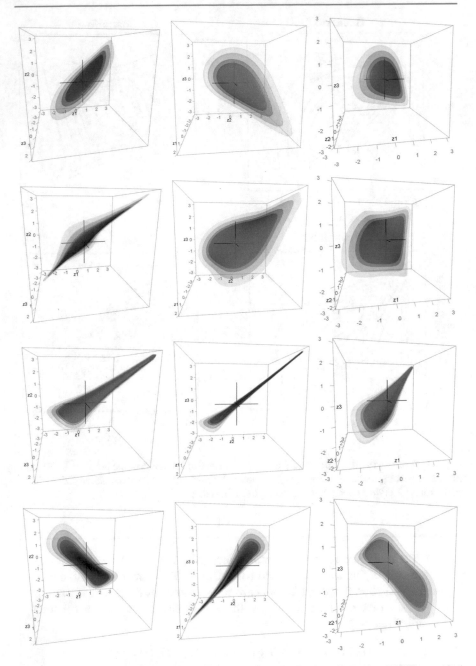

Fig. 4.2 `3Dcontours`: Three angles of 3D level surfaces for the three dimensional PCC's specified in Table 4.1 (top to bottom rows: Case 1 to Case 4)

Since

$$\frac{dv_1}{du_1} = \frac{dF_{1|2}^{-1}(u_1|x_2)}{du_1} = \frac{1}{f_{1|2}(F_{1|2}^{-1}(u_1|x_2)|x_2)}$$

and a similar expression for dv_3/du_3 holds it follows that

$$c_{13;2}(u_1, u_3; x_2) = \frac{f_{13|2}(F_{1|2}^{-1}(u_1|x_2), F_{3|2}^{-1}(u_3|x_2))}{f_{1|2}(F_{1|2}^{-1}(u_1|x_2)|x_2)f_{3|2}(F_{3|2}^{-1}(u_3|x_2)|x_2)}$$

$$= \frac{\dfrac{f_{123}(F_{1|2}^{-1}(u_1|x_2), x_2, F_{3|2}^{-1}(u_3|x_2))}{f_2(x_2)}}{\dfrac{f_{12}(F_{1|2}^{-1}(u_1|x_2), x_2)}{f_2(x_2)}\dfrac{f_{23}(x_2, F_{3|2}^{-1}(u_1|x_2))}{f_2(x_2)}} \qquad (4.12)$$

Using the density expression from Sklar's Theorem 1.9 gives

$$f_{123}(F_{1|2}^{-1}(u_1|x_2), x_2, F_{3|2}^{-1}(u_3|x_2)) = c_{123}(F_1(F_{1|2}^{-1}(u_1|x_2)), F_2(x_2), F_3(F_{3|2}^{-1}(u_3|x_2)))$$
$$\times f_1(F_{1|2}^{-1}(u_1|x_2))f_2(x_2)f_3(F_{3|2}^{-1}(u_3|x_2))$$
$$f_{12}(F_{1|2}^{-1}(u_1|x_2), x_2) = c_{12}(F_1(F_{1|2}^{-1}(u_1|x_2)), F_2(x_2))$$
$$\times f_1(F_{1|2}^{-1}(u_1|x_2))f_2(x_2)$$
$$f_{23}(x_2, F_{3|2}^{-1}(u_3|x_2)) = c_{23}(F_2(x_2), F_3(F_{3|2}^{-1}(u_3|x_2)))$$
$$\times f_3(F_{3|2}^{-1}(u_3|x_2))f_2(x_2)$$

Inserting this into (4.12) results in

$$c_{13;2}(u_1, u_3; x_2) = \frac{c_{123}(F_1(F_{1|2}^{-1}(u_1|x_2)), F_2(x_2), F_3(F_{3|2}^{-1}(u_3|x_2)))}{c_{12}(F_1(F_{1|2}^{-1}(u_1|x_2)), F_2(x_2))c_{23}(F_2(x_2), F_3(F_{3|2}^{-1}(u_3|x_2)))}$$
$$(4.13)$$

Ex 4.2 (Copula and density corresponding to (U_1, U_3) given $U_2 = u_2$) We
will denote the copula corresponding to the bivariate distribution function
$C_{13|2}(\cdot, \cdot|u_2)$ of (U_1, U_3) given $U_2 = u_2$ by $C_{13;2}^u(\cdot, \cdot|u_2)$ and its density by
$c_{13;2}^u(\cdot, \cdot|u_2)$, respectively. We will derive an expression for $c_{13;2}^u(\cdot, \cdot|u_2)$ and
relate it to $c_{13;2}(\cdot, \cdot|u_2)$ defined in (4.11). Using (1.18) we can express

$$c_{13;2}^u(u_1, u_3|u_2) = \frac{c_{13|2}(C_{1|2}^{-1}(u_1|u_2), C_{3|2}^{-1}(u_3|u_2))}{c_{1|2}(C_{1|2}^{-1}(u_1|u_2)|u_2)c_{3|2}(C_{3|2}^{-1}(u_3|u_2)|u_2)}$$

$$= \frac{c_{123}(C_{1|2}^{-1}(u_1|u_2), u_2, C_{3|2}^{-1}(u_2|u_2))}{c_{12}(C_{1|2}^{-1}(u_1|u_2), u_2)c_{23}(u_2, C_{3|2}^{-1}(u_3|u_2))} \tag{4.14}$$

Inverting the relationship between inverse conditional distribution functions as given in (1.28) it follows that

$$C_{1|2}^{-1}(u_1|F_2(x_2)) = F_1^{-1}(F_{1|2}^{-1}(u_1|x_2))$$
$$C_{3|2}^{-1}(u_3|F_2(x_2)) = F_3^{-1}(F_{3|2}^{-1}(u_3|x_2))$$

Let $u_2 = F_2(x_2)$ and inserting the above equations in (4.14) gives

$$c_{13;2}^{u}(u_1, u_3|u_2) = \frac{c_{123}(F_1(F_{1|2}^{-1}(u_1|x_2)), F_2(x_2), F_3(F_{3|2}^{-1}(u_3|x_2)))}{c_{12}(F_1(F_{1|2}^{-1}(u_1|x_2)), F_2(x_2))c_{23}(F_2(x_2), F_3(F_{3|2}^{-1}(u_3|x_2)))} \tag{4.15}$$

Comparing (4.15) with (4.13) shows that

$$c_{13;2}^{u}(u_1, u_3|u_2) = c_{13;2}(u_1, u_3|x_2) \text{ for } u_2 := F_2(x_2)$$

We conclude this section by illustrating the pair copula construction using the three dimensional red wine data of Example 1.9.

Ex 4.3 (WINE3: *Pair copula constructions*) In Example 3.4 we have already chosen pair copula families for the three pairs as given in Table 3.4. Now we investigate the three possible pair copula constructions given by (4.6), (4.7) and (4.8) for this data set with pseudo copula observation $(u_{i,acf}, u_{i,acc}, u_{i,acv})$ for $i = 1, \ldots, n$. Here we enumerated the variables as follows: $\mathtt{acf}{=}1$, $\mathtt{acc}{=}2$ and $\mathtt{acv}{=}3$. In particular we need to choose appropriate pair copula families for the distribution of

- **PCC1:** (U_{acf}, U_{acv}) given U_{acc} (corresponds to (4.6))
- **PCC2:** (U_{acf}, U_{acc}) given U_{acv} (corresponds to (4.7))
- **PCC3:** (U_{acv}, U_{acc}) given U_{acf} (corresponds to (4.8)),

where U_{acf}, U_{acv} and U_{acc} are the corresponding copula random variables of the variables $\mathtt{acf}$, $\mathtt{acv}$ and $\mathtt{acc}$. An approximate sample for the bivariate distribution (U_{acf}, U_{acv}) given U_{acc} is given by using the pseudo copula data

$$u_{i,acf|acc} := C_{acf|acc}(u_{i,acf}|u_{i,acc}; \hat{\theta}_{acf,acc}) \text{ and } u_{i,acv|acc} := C_{acv,acc}(u_{i,acv}|u_{i,acc}; \hat{\theta}_{acv,acc}),$$

Table 4.2 WINE3 : Three pair copula constructions together with empirically chosen pair copula families, their estimated Kendall's $\hat{\tau}$ and their estimated parameter $\hat{\theta}$

Pair copula	Copula family	$\hat{\tau}$	$\hat{\theta}$
PCC1			
(acf,acv)	Gumbel	0.48	1.94
(acv,acc)	Frank	−0.43	−4.57
(acf,acc;acv)	Gauss	0.13	0.20
PCC2			
(acf,acv)	Gumbel 270	−0.19	−1.23
(acv,acc)	Frank	−0.43	−4.57
(acf,acc;acv)	Gumbel	0.45	1.83
PCC3			
(acf,acv)	Gumbel 270	−0.19	−1.23
(acf,acc)	Gumbel	0.48	1.94
(acv,acc;acf)	Frank	−0.38	−3.95

where $\hat{\theta}_{acf,acc} = 1.9$, $\hat{\theta}_{acv,acc} = -4.6$ are the associated parameters of a 270° rotated Gumbel and a Frank copula, respectively. This bivariate sample can now be used to determine empirical normalized contour plots for $C_{acf,acv;acc}$ given in the left top panel of Fig. 4.3. The shape of the empirical normalized contours suggests to use a Gaussian copula for $C_{acf,acv;acc}$. The corresponding empirical Kendall's τ is used to estimate the copula parameter for the Gaussian copula. The corresponding fitted normalized Gaussian contours are given in the right top panel of Fig. 4.3.

The corresponding results for the other two pair copula constructions, i.e. for $C_{acf,acc;acv}$ is given in the middle panels of Fig. 4.3 and for $C_{acv,acc;acf}$ in the bottom panels of Fig. 4.3, respectively. For $C_{acf,acc;acv}$ the empirical marginally normalized contour plot suggested a Gumbel copula, while for $C_{acv,acc;acf}$ it is a Frank copula. The corresponding empirical Kendall's τ estimates are included in the right panel of the figures.

Table 4.2 summarizes our findings. In Chap. 7 we will discuss parameter estimation more formally.

Ex 4.4 (WINE3: *Three dimensional normalized contour plots*) We now illustrate the three dimensional PCC's estimated for the WINE3 data set using three dimensional normalized contour regions as developed and discussed in Killiches et al. (2017). The results are shown in Fig. 4.4. The first two columns

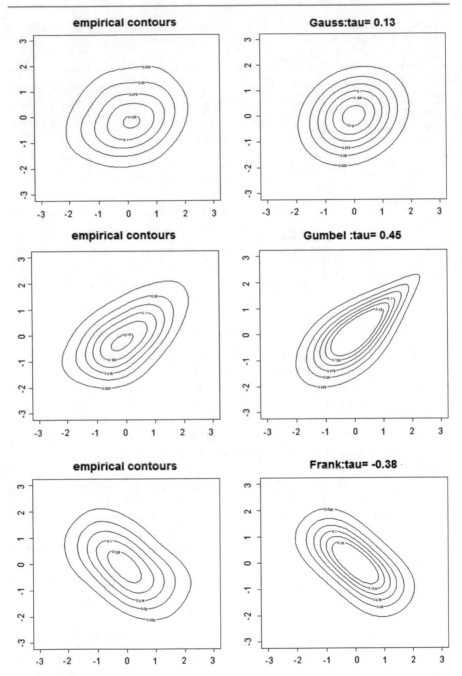

Fig. 4.3 WINE3 : Empirical (first column) based on pseudo copula data and normalized contour plots (second column) for a Gaussian conditional pair copula $C_{acf,acv;acc}$ needed for PCC1 (top right), a 270° rotated Gumbel copula $C_{acf,acc;acv}$ needed for PCC2 (middle right) and a Frank copula $C_{acv,acc;acf}$ needed for PCC3 (bottom right)

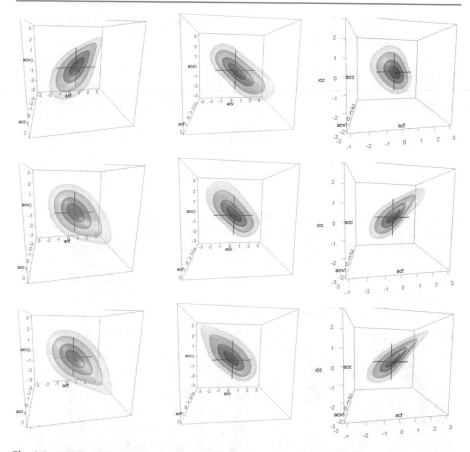

Fig. 4.4 WINE3 : Three dimensional normalized contour plots corresponding to the fitted PCC's in Table 4.2 (top row: PCC1, middle row: PCC2 and bottom row: PCC3)

reflect the unconditional pair copula choices, while the last column corresponds to an unconditional view of the last pair, which was not modeled directly. Thus this pair is only obtainable by integration of the full density over the conditioning variable.

4.2 Pair-Copula Constructions of Drawable D-vine and Canonical C-vine Distributions

The starting point is again the decomposition of a multivariate density into products of conditional densities. For this let $(X_1, ..., X_d)$ be a set of variables with joint distribution $F_{1,...,d}$ and density $f_{1,...,d}$, respectively. Consider the decomposition

$$f_{1,...,d}(x_1, ..., x_d) = f_{d|1,...,d-1}(x_d|x_1, \cdots, x_{d-1}) f_{1...,d-1}(x_1, \cdots, x_{d-1})$$

$$= \cdots = \left[\prod_{t=2}^{d} f_{t|1,...,t-1}(x_t|x_1, \cdots, x_{t-1}) \right] \times f_1(x_1). \qquad (4.16)$$

Here $f(\cdot|\cdot)$ and later $F(\cdot|\cdot)$ denote conditional densities and cumulative distribution functions of the variables indicated by the subscripts, respectively.

As a second ingredient we use again Sklar's Theorem 1.9 for dimension $d = 2$ and consequently Lemma 1.15 to expand the conditional densities. In the following we need the notion of copulas associated with bivariate conditional distributions in contrast to bivariate conditional distributions on the copula scale. Therefore we introduce some useful abbreviations to shorten notation.

Definition 4.5 (*Copulas associated with bivariate conditional distributions*) Let $(X_1, ..., X_d)$ be a set of random variables.

- Let D be a set of indices from $\{1, ..., d\}$ not including i and j. The *copula associated with the bivariate conditional distribution* (X_i, X_j) *given that* $X_D = x_D$ is denoted by $C_{ij;D}(\cdot, \cdot; x_D)$.
- In contrast the *conditional distribution function of* (U_i, U_j) *given* $U_D = u_D$ is expressed as $C_{ij|D}(\cdot, \cdot; u_D)$ with bivariate density function $c_{ij|D}(\cdot, \cdot; u_D)$.
- For distinct indices i, j and $D := \{i_1, ..., i_k\}$ with $i < j$ and $i_1 < \cdots < i_k$ we use the abbreviation

$$c_{i,j;D} := c_{i,j;D}(F_{i|D}(x_i|x_D), F_{j|D}(x_j|x_D); x_D). \qquad (4.17)$$

Remark 4.6 (*Difference between* $C_{ij;D}(\cdot, \cdot; x_D)$ *and* $C_{ij|D}(\cdot, \cdot; u_D)$) In general the copula $C_{ij;D}(\cdot, \cdot; x_D)$ is not equal to the bivariate distribution function $C_{ij|D}(\cdot, \cdot; u_D)$ even if $u_D = F_D(x_D)$ holds, since $C_{ij|D}$ might have non uniform margins, while $C_{ij;D}$ needs to have uniform margins. In Example 3.5 we have explored these differences in the case of the trivariate Frank copula.

These notations now allow us to decompose the joint density as follows.

Theorem 4.7 (Drawable vine (D-vine) density) *Every joint density $f_{1,...,d}$ can be decomposed as*

$$f_{1,...,d}(x_1, \ldots, x_d) = \left[\prod_{j=1}^{d-1}\prod_{i=1}^{d-j} c_{i,(i+j);(i+1)...,(i+j-1)}\right] \cdot \left[\prod_{k=1}^{d} f_k(x_k)\right],$$

(4.18)

where we used the abbreviation introduced in Eq. (4.17). The distribution associated with this density decomposition is called a drawable vine (D-vine).

Proof Using Lemma (1.15) for the conditional distribution of (X_1, X_t) given $X_2, \ldots X_{t-1}$ we can express the conditional density $f_{t|1,...,t-1}(x_t|x_1, \ldots, x_{t-1})$ recursively as

$$f_{t|1,...,t-1}(x_t|x_1, \ldots, x_{t-1}) = c_{1,t|2,...,t-1} \times f_{t|2,...,t-1}(x_t|x_2, \ldots, x_{t-1})$$

$$= \left[\prod_{s=1}^{t-2} c_{s,t;s+1,...,t-1}\right] \times c_{(t-1),t} \times f_t(x_t) \quad (4.19)$$

Using (4.19) in decomposition (4.16) and setting $s = i, t = i + j$ gives

$$f_{1,...,d}(x_1, \ldots, x_d) = \left[\prod_{t=2}^{d}\prod_{s=1}^{t-2} c_{s,t;s+1,...,t-1}\right] \cdot \left[\prod_{t=2}^{d} c_{(t-1),t}\right]\left[\prod_{k=1}^{d} f_k(x_k)\right]$$

$$= \left[\prod_{j=1}^{d-1}\prod_{i=1}^{d-j} c_{i,(i+j);(i+1)...,(i+j-1)}\right] \cdot \left[\prod_{k=1}^{d} f_k(x_k)\right].$$

Note that the decomposition (4.18) of the joint density consists of pair-copula densities $c_{i,j|D}(\cdot, \cdot; x_D)$ evaluated at conditional distribution functions $F_{i|D}(x_i|x_{i_1}, \cdots, x_{i_k})$ and $F_{j|D}(x_j|x_{i_1}, \cdots, x_{i_k})$ for specified indices $i, j, i_1, \cdots, i_k$ and marginal densities f_k. This is the reason why we call such a decomposition a *pair-copula decomposition*. This class of decompositions was named by Bedford and Cooke (2001, 2002) the class of *D-vine distributions*.

A second class of decompositions is possible, when we apply Lemma (1.15) to the conditional distribution of (X_{t-1}, X_t) given $X_1, \cdots, X_{t-2}$ to express $f_{t|1,...,t-1}(x_t|x_1, \cdots, x_{t-1})$ recursively. This second class is given in Theorem 4.8.

Theorem 4.8 (Canonical vine (C-vine) density) *The joint density can be decomposed as*

$$f(x_1, ..., x_d) = \left[\prod_{j=1}^{d-1}\prod_{i=1}^{d-j} c_{j,j+i;1,\cdots,j-1}\right] \times \left[\prod_{k=1}^{d} f_k(x_k)\right],$$

(4.20)

we used the abbreviation introduced in Eq. (4.17). According to Bedford and Cooke (2001, 2002) this PCC is called a canonical vine (C-vine) distribution.

Proof Exercise.

We now extend the simplifying assumption of Definition 4.2 to C- and D-vines.

Definition 4.9 (*Simplifying assumption for C and D-vines*) If

$$c_{ij,D}(F_{i|D}(x_i|x_D), F_{j|D}(x_j|x_D); x_D) = c_{ij,D}(F_{i|D}(x_i|x_D), F_{j|D}(x_j|x_D))$$

holds for all x_D and i, j and D are chosen to occur in (4.18)–(4.20), then the corresponding D-vine (C-vine) distribution is called *simplified*.

For illustration we specify simplified C- and D-vines in four dimensions.

Ex 4.5 (*Simplified dimensional D- and C-vine density in four dimensions*) For $d = 4$ the simplified D-vine density has the following form

$$f_{1234}(x_1, x_2, x_3, x_4) = [\prod_{i=1}^{4} f_i(x_i)] \times c_{12}(x_1, x_2) \times c_{23}(x_2, x_3) \times c_{34}(x_3, x_4)$$
$$\times c_{13;2}(F_{1|2}(x_1|x_2), F_{3|2}(x_3|x_2)) \times c_{24;3}(F_{2|3}(x_2|x_3), F_{4|3}(x_4|x_3))$$
$$\times c_{14;23}(F_{1|23}(x_1|x_2, x_3), F_{4|23}(x_4|x_2, x_3))$$

For the simplified C-vine we have

$$f_{1234}(x_1, x_2, x_3, x_4) = [\prod_{i=1}^{4} f_i(x_i)] \times c_{12}(x_1, x_2) \times c_{13}(x_1, x_3) \times c_{14}(x_1, x_4)$$
$$\times c_{23;1}(F_{2|1}(x_2|x_1), F_{3|1}(x_3|x_1)) \cdot c_{24;1}(F_{2|1}(x_2|x_1), F_{4|1}(x_4|x_1))$$
$$\times c_{34;12}(F_{3|12}(x_3|x_1, x_2), F_{4|12}(x_4|x_1, x_2)).$$

4.3 Conditional Distribution Functions Associated with Multivariate Distributions

The density of D- and C-vines requires the evaluation of conditional distribution functions. We discuss these evaluations now in the context of a general distribution. We begin with an illustrative example.

Ex 4.6 (Computation of $F_{1|23}$ and $C_{1|23}$ in the simplified and non simplified case) Using Sklar's theorem for the bivariate conditional density $f_{13|2}$ and Lemma 1.15 to derive the associated marginal density $f_{1|23}$ we have in the simplified case

$$F_{1|23}(x_1|x_2, x_3) = \int_{-\infty}^{x_1} f_{1|23}(y_1|x_2, x_3)dy_1$$

$$= \int_{-\infty}^{x_1} c_{13;2}(F_{1|2}(y_1|x_2), F_{3|2}(x_3|x_2))f_{1|2}(y_1|x_2)dy_1$$

$$= \int_{-\infty}^{x_1} \frac{\partial}{\partial F_{1|2}(y_1|x_2)} \frac{\partial}{\partial F_{3|2}(x_3|x_2)} C_{13;2}(F_{1|2}(y_1|x_2), F_{3|2}(x_3|x_2))f_{1|2}(y_1|x_2)dy_1$$

$$= \frac{\partial}{\partial F_{3|2}(x_3|x_2)} \int_{-\infty}^{x_1} \left[\frac{\partial}{\partial y_1} C_{13;2}(F_{1|2}(y_1|x_2), F_{3|2}(x_3|x_2)) \right] dy_1$$

$$= \frac{\partial}{\partial F_{3|2}(x_3|x_2)} C_{13;2}(F_{1|2}(x_1|x_2), F_{3|2}(x_3|x_2)).$$

In the copula case we have for $C_{1|23}$ similarly

$$C_{1|23}(u_1|u_2, u_3) = \frac{\partial}{\partial C_{3|2}(u_3|u_2)} C_{13;2}(C_{1|2}(u_1|u_2), C_{3|2}(u_3|u_2))$$

$$= \frac{\partial}{\partial v_2} C_{13;2}(h_{1|2}(u_1|u_2), v_2))|_{v_2 = h_{3|2}(u_3|u_2)}$$

$$= h_{1|3;2}(h_{1|2}(u_1|u_2)|h_{3|2}(u_3|u_2)), \tag{4.21}$$

where

$$h_{1|3;2}(v_1|v_3) := \frac{\partial}{\partial v_3} C_{13;2}(v_1, v_3)$$

is the h function with respect to the first argument v_1 of the bivariate copula $C_{13;2}$ (compare to Definition 1.16). In particular $h_{1|3;2}(\cdot|v_3)$ is the conditional distribution function of V_1 given $V_3 = v_3$ when the joint distribution of (V_1, V_3) is given by $C_{13;2}$. Here we used the fact that we are in the simplified case, i.e. $h_{1|3;2}$ does not depend on the conditioning value u_2.

In the case of a non simplified vine the copula density $c_{13;2}$ depends on u_2 and therefore we define

$$h_{1|3;2}(v_1|v_3; u_2) := \frac{\partial}{\partial v_3} C_{13;2}(v_1, v_3; u_2)$$

and can compute $C_{1|23}$ accordingly.

This last example can now be generalized and we have the following result.

Theorem 4.10 (Recursion for conditional distribution functions) *Let X be a random variable and $\mathbf{Y}$ be a random vector which have an absolutely continuous joint distribution. Let Y_j a component of $\mathbf{Y}$ and denote the sub-vector of $\mathbf{Y}$ with Y_j removed by $\mathbf{Y}_{-j}$. In this case the conditional $F_{X|\mathbf{Y}}(\cdot|\mathbf{y})$ distribution of X given $\mathbf{Y} = \mathbf{y}$ satisfies the following recursion*

$$F_{X|\mathbf{Y}}(\cdot|\mathbf{y}) = \frac{\partial C_{X,Y_j;\mathbf{Y}_{-j}}(F_{X|Y_j}(x|\mathbf{y}_{-j}), F_{Y_j|Y_j}(y|\mathbf{y}_{-j}))}{\partial F_{Y_j|\mathbf{Y}_{-j}}(y_j|\mathbf{y}_{-j})},$$

where $C_{X,Y_j;\mathbf{Y}_{-j}}(\cdot, \cdot|\mathbf{y}_{-j})$ denotes the copula corresponding to (X, Y_j) given $\mathbf{Y}_{-j} = \mathbf{y}_{-j}$.

Proof This result follows directly from the chain rule of differentiation and was first stated in Joe (1996).   □

4.4 Exercises

Exer 4.1
Exploratory copula choices for the three dimensional uranium *data*: Consider as in Example 3.4 the three dimensional subset of the uranium data set contained in the R package copula with variables Cobalt (Co), Titanium (Ti) and Scandium (Sc).

- Use the copula choices you made for the pairs of variables to construct the pseudo copula data needed for the estimation of the three conditional pair copulas $C_{Co,Ti;Sc}$, $C_{Co,Sc;Ti}$ and $C_{Sc,Ti;Co}$.
- Construct the corresponding empirical normalized contour plots and suggest appropriate pair copula families.
- Use these choices of pair copula families to construct a table (compare to Table 4.2) of empirical Kendall's τ and θ estimates using inversion for the three possible pair copula construction.

Exer 4.2
C-vine decomposition: Prove Theorem 4.8.

Exer 4.3
Gaussian pair copula construction in 3 dimensions: Show that in a pair copula decomposition of the three dimensional normal distribution $N_3(\mu, \Sigma)$ the copula

density associated with the conditional distribution (X_1, X_3) given $X_2 = x_2$ does not depend on x_2. This means that the simplifying assumption is satisfied.

Exer 4.4

Pair copula construction in 3 dimensions of the multivariate Student t distribution: Show that in a pair copula decomposition of a three dimensional Student t distribution $t_3(\mu, \nu, \Sigma)$ the copula density associated with the the conditional distribution (X_1, X_3) given $X_2 = x_2$ does not depend on x_2. This means that the simplifying assumption is satisfied.

Exer 4.5

Pair copula construction in 3 dimensions of the multivariate Clayton copula: For the three dimensional Clayton copula (see Eaxmple 3.6) defined as

$$C(u_1, u_2, u_3) = (u_1^{-\delta} + u_2^{-\delta} + u_3^{-\delta} - 2)^{\frac{1}{\delta}}$$

for $0 < \delta < \infty$ derive the following properties:

- The bivariate conditional density $c_{13|2}$ depends on the conditioning value u_2.
- The copula density associated with the the conditional distribution (U_1, U_3) given $U_2 = u_2$ does not depend on u_2 and agrees with by a bivariate Clayton copula density with parameter $\frac{\delta}{1+\delta}$. This means that the simplifying assumption is satisfied.

The Clayton copula in d dimensions has been considered in Joe (2014) in Sect. 4.6.2.

Exer 4.6

Conditional distributions in four dimensional simplified C-vines: Show that the required conditional distributions in the simplified C-vine in four dimensions introduced in Example 4.6 can be computed using only the pair copulas specified in the construction. This holds true for C- and D-vines in arbitrary dimension.

Regular Vines

5

In the last chapter we saw that we can construct trivariate distributions using only bivariate building blocks. Additionally we introduced special vine distribution classes such as C- and D-vine distributions in arbitrary dimensions through recursive conditioning. In this chapter we generalize this construction principle allowing for different conditioning orders. As we already noted the construction is not unique, therefore it will be important to allow for different constructions and to organize them. We present now the approach developed by Bedford and Cooke (2001, 2002) and presented in Kurowicka and Cooke (2006), Stöber and Czado (2017).

5.1 Necessary Graph Theoretic Background

Since different constructions are identified with graph theoretical concepts we first present the necessary background from graph theory. A good resource is Diestel (2006).

Definition 5.1 (*Graph, node, edge, degree*) We use the following notation.

- A *graph* is a pair $G = (N, E)$ of sets such that $E \subseteq \{\{x, y\} : x, y \in N\}$.
- The elements of E are called *edges* of the graph G, while the elements of N are called *nodes*.
- The number of neighbors of a node $v \in N$ is the *degree* of v, denoted by $d(v)$.

© Springer Nature Switzerland AG 2019
C. Czado, *Analyzing Dependent Data with Vine Copulas*, Lecture Notes in Statistics 222, https://doi.org/10.1007/978-3-030-13785-4_5

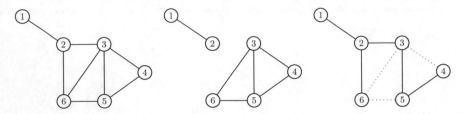

Fig. 5.1 Example graphs (I): Left panel: Graph for Example 5.1. Middle panel: disconnected graph with components on nodes $N_1 = \{1, 2\}$ and $N_2 = \{3, 4, 5, 6\}$. Right panel: spanning tree of the graph in the left panel

---

Ex 5.1 (*Example of a graph*)   Consider the graph given in the left panel of Fig. 5.1. It is a graph with $G = (N, E)$ with nodes $N = \{1, 2, 3, 4, 5, 6\}$ and edges

$$E = \{\{1, 2\}, \{2, 3\}, \{2, 6\}, \{3, 4\}, \{3, 5\}, \{3, 6\}, \{4, 5\}, \{5, 6\}\}.$$

For example the *degree* of node 3 is $d(3) = 4$.

The graph G defined in Definition 5.1 is usually referred to as *undirected*, since the order of nodes corresponding to an edge is arbitrary. In a *directed* graph the order of the edge matters, therefore edge set E is tuple with two elements and is denoted by $E \subseteq \{(x, y) : x, y \in N\}$. If there is a function $w : E \to \mathbb{R}$, then the graph G is called *weighted* and denoted by $G = (N, E, w)$, i.e., weights are assigned to each edge. Moreover, if $E = \{\{x, y\} : x, y \in N\}$ in Definition 5.1, then G is called *complete*.

Ex 5.2 (*Illustration of basic graph concepts*)   The left panel of Fig. 5.2 shows a directed graph, while the middle panel gives a complete graph on four nodes.

A *subgraph* of a graph $G = (N, E)$ is a graph $G' = (N', E')$ with node set $N' \subseteq N$ and edge set $E' \subseteq E$. Important examples of graphs are *paths* and *cycles*, which often occur as subgraphs of interest.

Definition 5.2 (*Path, cycle*) A *path* is a graph $P = (N, E)$ with node set $N = \{v_0, v_1, ..., v_k\}$ and edges $E = \{\{v_0, v_1\}, \{v_1, v_2\}, ..., \{v_{k-1}, v_k\}\}$. A *cycle* is a path with $v_0 = v_k$.

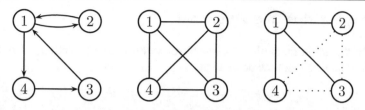

Fig. 5.2 Example graphs (II): left panel: directed graph. Middle panel: complete graph on four nodes. Right panel: spanning star with root node 1 in a complete graph

For example the graph in the left panel of Fig. 5.1 has a cycle given by $2 - 3 - 5 - 6 - 2$ and a path by $1 - 2 - 3 - 4 - 5 - 6$.

A graph G is called *connected* if any two of its nodes are linked by a path in G. A disconnected graph is shown in the middle panel of Fig. 5.1. Further, a path in G containing every node of G is called a *Hamiltonian path*,. The path given by $1 - 2 - 3 - 4 - 5 - 6$ in the left panel of Fig. 5.1 is a Hamiltonian path. A cycle which contains every node of G is called a *Hamiltonian cycle*. Two nodes are *adjacent* if there is an edge connecting them, otherwise they are called *non-adjacent*.

The most important class of graphs that will be considered in the following are *trees*, which are connected and do not contain cycles. They can be characterized by the following theorem, where $G \pm e$ denotes a graph with an added/removed edge e.

Theorem 5.3 (Characterization of trees) *The following statements are equivalent for a graph $T = (N, E)$:*

1. *T is a tree.*
2. *Any two nodes of T are connected by a unique path in T.*
3. *T is minimally connected, i.e., T is connected but $T - e$ is disconnected for every edge $e \in E$.*
4. *T is maximally acyclic, i.e., T contains no cycle but $T + \{x, y\}$ does for any two non-adjacent nodes $x, y \in N$.*

A *spanning tree* of a graph $G = (N, E)$ is a subgraph $T = (N_T, E_T)$, which is a tree with $N_T = N$. Moreover, a tree, which has a node v_0 with $d(v_0) = |N| - 1$, will be called a *star* and v_0 the *root node*. Obviously, in star trees it holds that $d(v) = 1 \ \forall v \in N \setminus \{v_0\}$.

The right panel of Fig. 5.1 shows a spanning tree of the graph in the left panel. A spanning star with root node 1 is shown in the right panel of Fig. 5.2.

5.2   Regular Vine Tree Sequences

Based on the graph theoretic concepts introduced in the last section, a regular (R-) vine tree sequence can now be defined.

Definition 5.4 (*Regular (R-) vine tree sequence*)   The set of trees $\mathcal{V} = (T_1, \ldots, T_{d-1})$ is a *regular vine tree sequence* on d elements if:

(1) Each tree $T_j = (N_j, E_j)$ is connected, i.e. for all nodes $a, b \in T_j$, $j = 1, \ldots, d - 1$, there exists a path $n_1, \ldots, n_k \subset N_j$ with $a = n_1, b = n_k$.
(2) T_1 is a tree with node set $N_1 = \{1, \ldots, d\}$ and edge set E_1.
(3) For $j \geq 2$, T_j is a tree with node set $N_j = E_{j-1}$ and edge set E_j.
(4) For $j = 2, \ldots, d - 1$ and $\{a, b\} \in E_j$ it must hold that $|a \cap b| = 1$.

Remark 5.5 (*Proximity condition*)   The property (4) is called the *proximity condition*. It ensures that if there is an edge e connecting a and b in tree T_j, $j \geq 2$, then a and b (which are edges in T_{j-1}) must share a common node in T_{j-1}.

Ex 5.3 (*Six dimensional regular vine tree sequence specified using the set notation*)   We consider now an example for a vine tree sequence on six nodes with trees T_1 until T_5 by specifying their node sets N_i and edge sets E_i for $i = 1, \ldots, 5$ as follows:

T_1 $N_1 = \{1, 2, 3, 4, 5, 6\}$
$\quad\quad E_1 = \{\{12\}, \{13\}, \{34\}, \{15\}, \{56\}\}$

T_2 $N_2 = E_1 = \{\{12\}, \{13\}, \{34\}, \{15\}, \{56\}\}$
$\quad\quad E_2 = \{\{\{12\}, \{13\}\}, \{\{13\}, \{34\}\}, \{\{13\}, \{15\}\}, \{\{15\}, \{56\}\}\}$

T_3 $N_3 = E_2 = \{\{\{12\}, \{13\}\}, \{\{13\}, \{34\}\}, \{\{13\}, \{15\}\}, \{\{15\}, \{56\}\}\}$
$\quad\quad E_3 = \{\{\{\{12\}, \{13\}\}, \{\{13\}, \{34\}\}\},$
$\quad\quad\quad \{\{\{13\}, \{34\}\}, \{\{13\}, \{15\}\}\},$
$\quad\quad\quad \{\{\{13\}, \{15\}\}, \{\{15\}, \{16\}\}\}\}$

T_4 $N_4 = E_3 = \{\{\{\{12\}, \{13\}\}, \{\{13\}, \{34\}\}\},$
$\quad\quad\quad\quad\quad \{\{\{13\}, \{34\}\}, \{\{13\}, \{15\}\}\},$
$\quad\quad\quad\quad\quad \{\{\{13\}, \{15\}\}, \{\{15\}, \{16\}\}\}\}$
$\quad\quad E_4 = \{\{\{\{\{12\}, \{13\}\}, \{\{13\}, \{34\}\}\}, \{\{\{13\}, \{34\}\}, \{\{13\}, \{15\}\}\}\},$
$\quad\quad\quad \{\{\{\{13\}, \{34\}\}, \{\{13\}, \{15\}\}\}, \{\{\{13\}, \{15\}\}, \{\{15\}, \{16\}\}\}\}\}$

T_5 $N_5 = E_4$
$\quad\quad E_5 = \{\{\{\{\{\{12\}, \{13\}\}, \{\{13\}, \{34\}\}\}, \{\{\{13\}, \{34\}\}, \{\{13\}, \{15\}\}\}\},$
$\quad\quad\quad \{\{\{\{13\}, \{34\}\}, \{\{13\}, \{15\}\}\}, \{\{\{13\}, \{15\}\}, \{\{15\}, \{16\}\}\}\}\}\}$.

To shorten notation we used the abreviation ij for the unordered tuple $\{i, j\}$. You can easily check that the proximity condition given in Remark 5.5 holds for this example. As an example for a pair of nodes in tree T_2 that cannot be directly joined by an edge, consider the pair $\{1, 2\}$ and $\{3, 4\}$ from the node set N_2 in T_2. They cannot be connected, since they do not share as edges in E_1 a common node in T_1.

From Example 5.3 we see that the set notation becomes unmanageable even for small numbers of nodes, therefore we now introduce a simplified edge notation for a regular vine tree sequence. It will also be used later to assign bivariate copulas for the regular vine construction of multivariate distributions. For this some further notation is needed.

Definition 5.6 (*m-child and m-descendent*) For node e we define

- If node e is an element of node f, then we call f an *m-child* of e.
- Conversely if e is reachable from f via the membership relationship $e \in e_1 \in \dots \in f$, then e is called an *m-descendent* of f.

Ex 5.4 (*m descendents and m-children in Example* 5.3) Table 5.1 lists the associated m-descendents and m-children for the R-vine tree structure of Example 5.3 involving the first 4 nodes of tree T_1.

Definition 5.7 (*Complete union and conditioned sets*) For any edge $e \in E_i$ define the set

$$A_e := \left\{ j \in N_1 | \exists \, e_1 \in E_1, \dots, e_{i-1} \in E_{i-1} \text{ such that } j \in e_1 \in \dots \in e_{i-1} \in e \right\}.$$

The set A_e is called the *complete union* of the edge e. The *conditioning set* D_e of an edge $e = \{a, b\}$ is defined by

$$D_e := A_a \cap A_b$$

Table 5.1 Children and descendents: m-children and m-descendents in the R-vine tree sequence of Example 5.3 involving the first 4 nodes of tree T_1

Node e	m-children of e	m-descendents of $e \in N_1$
1	–	1
2	–	2
3	–	3
{12}	{{12}, {23}}	1, 2
{23}	{{12}, {23}}, {{23}, {34}}	2, 3
{34}	{{23}, {34}}	3, 4
{{12}, {23}}	{{{12}, {23}}, {{23}, {34}}}	1, 2, 3
{{23}, {34}}	{{{12}, {23}}, {{23}, {34}}}	1, 2, 3
{{{12}, {23}}, {{23}, {34}}}	–	1, 2, 3, 4

and the *conditioned sets* $C_{e,a}$ and $C_{e,b}$ are given by

$$C_{e,a} := A_a \setminus D_e, C_{e,b} := A_b \setminus D_e \text{ and } C_e := C_{e,a} \cup C_{e,b}.$$

We often abbreviate each edge $e = (C_{e,a}, C_{e,b}; D_e)$ in the vine tree sequence by

$$e = (e_a, e_b; D_e). \tag{5.1}$$

Note that the complete union A_e is the subset of $\{1, ..., d\}$ which consists of the m-descendents of e. Further the conditioned sets $C_{e,a}$ and $C_{e,b}$ are singletons and the size of a conditioning set of an edge in tree T_j is $j - 1$ for $j = 2, ..., j - 1$ as proven in Chap. 4 of Kurowicka and Cooke (2006). We use a semi column to separate the conditioned set from the conditioning set, since later we link edges to bivariate copulas associated with conditional distributions. This yields a consistent notation with Definition 4.5.

Ex 5.5 (Complete union, conditioned and conditioning sets) We illustrate these sets using the edge $e = \{\{1, 2\}, \{1, 3\}\}$ of Example 5.3. The complete union contains all nodes in tree T_1, which are included in e, i.e. $A_e = \{1, 2, 3\}$. The conditioning set of e is given by $D_e = \{1\}$, since $A_a = \{1, 2\}$ and $A_b = \{1, 3\}$. The conditioned set of edge e is therefore $C_e = \{2, 3\}$.

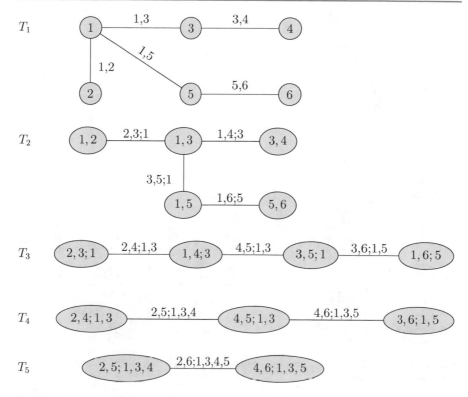

Fig. 5.3 Regular vine tree sequence: The six dimensional regular vine tree sequence of Example 5.3, from the lowest tree T_1 to the highest tree T_5 with edge labels based given by (5.1)

Ex 5.6 (Six dimensional regular vine tree structure using conditioned sets and conditioning sets) We consider again the regular vine tree sequence of Example 5.3 but this time with the edges labeled by their conditioned sets separated by a semi colon from their conditioning set as introduced in (5.1). The resulting tree sequence with edge labels is given in Fig. 5.3.

Remark 5.8 (Number of regular vine tree structures) Counting all possibilities of choosing edges, Morales-Nápoles (2011) shows that there are $(d!/2) \cdot 2^{\binom{d-2}{2}}$ R-vine tree sequence in d dimensions.

We now consider two important sub classes of R-vine tree sequences.

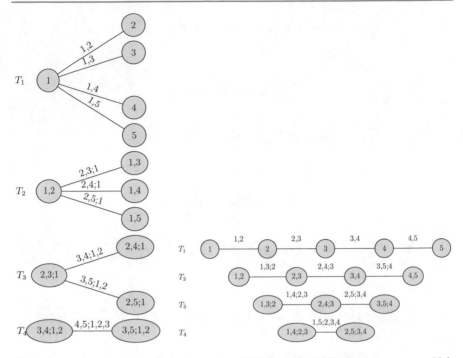

Fig. 5.4 C- and D-vine: C-vine tree sequence (left panel) and D-vine tree sequence (right panel) in four our dimensions

Definition 5.9 (*C-vine tree sequence, D-vine tree sequence*) A regular vine tree sequence $\mathcal{V} = (T_1, \ldots, T_{d-1})$ is called

- *D-vine tree sequence* if for each node $n \in N_i$ we have $|\{e \in E_i | n \in e\}| \leq 2$.
- *C-vine tree sequence* if in each Tree T_i there is one node $n \in N_i$ such that $|\{e \in E_i | n \in e\}| = d - i$. Such a node is called the *root node* of tree T_i.

Ex 5.7 (*C- and D-vine tree sequences in five dimensions*) The vine tree sequences of the C- and D-vine in five dimensions are given in Fig. 5.4, respectively.

Remark 5.10 (*Implications of the proximity condition on C- and D-vine tree sequences*) For a D-vine tree sequence the proximity condition of Definition 5.5 induces that once tree T_1 is fixed all other trees T_2 to T_{d-1} are determined. For a

C-vine tree sequence the proximity condition allows to choose $d - i + 1$ different root nodes in Tree T_i for $i = 1, \ldots, d - 1$.

Note that the tree structure of a D-vine resembles a grape vine, therefore (Bedford and Cooke 2001) called the linked tree sequence of Definition 5.4 a vine.

5.3   Regular Vine Distributions and Copulas

Up to now the R-vine tree sequence of Definition 5.4 is only a graph theoretic object, which does not contain any stochastic component. We now add such a component.

Definition 5.11 (*Regular vine distribution*) The joint distribution F for the d dimensional random vector $X = (X_1, \ldots, X_d)$ has a *regular vine distribution*, if we can specify a triplet $(\mathcal{F}, \mathcal{V}, \mathcal{B})$ such that:

1. **Marginal distributions**: $\mathcal{F} = (F_1, \ldots, F_d)$ is a vector of continuous invertible marginal distribution functions, representing the marginal distribution functions of the random variable X_i, $i = 1, \ldots, d$.
2. **Regular vine tree sequence**: $\mathcal{V}$ is an R-vine tree sequence on d elements.
3. **Bivariate copulas**: The set $\mathcal{B} = \{C_e | e \in E_i; i = 1, \ldots, d - 1\}$, where C_e is a symmetric bivariate copula with density. Here E_i is the edge set of tree T_i in the R-vine tree sequence $\mathcal{V}$.
4. **Relationship between R-vine tree sequence $\mathcal{V}$ and the set $\mathcal{B}$ of bivariate copulas**: For each $e \in E_i$, $i = 1, \ldots, d - 1$, $e = \{a, b\}$, C_e is the copula associated with the conditional distribution of $X_{C_{e,a}}$ and $X_{C_{e,b}}$ given $\mathbf{X}_{D_e} = \mathbf{x}_{D_e}$. Further $C_e(., .)$ does not depend on the specific value of $\mathbf{x}_{D_e}$.

Remark 5.12 (*Simplifying assumption for regular vine distributions*) The assumption in Definition 5.11 that the bivariate copulas $C_e(., .)$ do not depend on the specific value of $\mathbf{x}_{D_e}$ is called the *simplifying assumption*.

Ex 5.8 (*Example* 5.3 *continued*) The copula C_e for edge $e = \{\{1, 3\}, \{3, 4\}\}$ of Example 5.3 indicates that the conditional bivariate distribution function of (X_1, X_4) given $X_1 = x_1$ is given by

$$F_{14|3}(x_1, x_4 | x_3) = C_e\left(F_{1|3}(x_1 | x_3), F_{4|3}(x_4 | x_3)\right).$$

Definition 5.13 (*Pair copula and copula density associated with edge e*) We will denote the copula C_e corresponding to edge e by $C_{C_{e,a}C_{e,b};D_e}$ and the corresponding density by $c_{C_{e,a}C_{e,b};D_e}$, respectively. This copula is also called a *pair copula*.

Remark 5.14 (*Non-symmetric pair copulas in an regular vine distribution*) Since we define the regular vine tree sequence as a set of undirected graphs in Definition 5.4, we had to restrict ourselves to symmetric bivariate copulas (see Remark 3.10 for a definition of a symmetric copula) in Definition 5.11. This is only a formal restriction, since the theory presented in this chapter remains also true, when directed edges are used. This allows us to identify non-symmetric pair copulas. However symmetry is a common assumption in the literature on vines in order to be able to use the set notation with unordered elements as illustrated in Example 5.3. In simulations and applications we will indicate how to work with non-symmetric copulas.

Bedford and Cooke (2002) showed that the *R-vine triplet* $(\mathcal{F}, \mathcal{V}, \mathcal{B})$ with properties (1)–(3) of Definition 5.11 can be uniquely connected to a d dimensional distribution F, in particular the following theorem holds.

Theorem 5.15 (Existence of a regular vine distribution) *Assume that* $(\mathcal{F}, \mathcal{V}, \mathcal{B})$ *satisfy the properties (1)–(3) of Definition 5.11, then there is a unique d dimensional distribution F with density*

$$f_{1,\ldots d}(x_1,\ldots x_d) = f_1(x_1) \cdot \ldots \cdot f_d(x_d) \tag{5.2}$$

$$\cdot \prod_{i=1}^{d-1} \prod_{e \in E_i} c_{C_{e,a}C_{e,b};D_e}(F_{C_{e,a}|D_e}(x_{C_{e,a}}|x_{D_e}), F_{C_{e,b}|D_e}(x_{C_{e,b}}|x_{d_e})),$$

such that for each $e \in E_i$, $i = 1,\ldots, d-1$, *with* $e = \{a,b\}$ *we have for the distribution function of* $X_{C_{e,a}}$ *and* $X_{C_{e,b}}$ *given* $X_{D_e} = x_{D_e}$

$$F_{C_{e,a}C_{e,b}|D_e}\left(x_{C_{e,a}}, x_{C_{e,b}}|x_{D_e}\right) = C_e\left(F_{C_{e,a}|D_e}(x_{C_{e,a}}|x_{D_e}), F_{C_{e,b}|D_e}(x_{C_{e,b}}|x_{D_e})\right).$$

Further the one dimensional margins of F are given by $F_i(x_i)$, $i = 1,\ldots, d$.

Remark 5.16 (Extended vine tree sequences) Bedford and Cooke (2001) extended the regular vine tree sequence to Cantor trees, which then allows to proof the existence of the regular vine distributions more elegantly.

Theorem 5.15 allows to specify regular vine distributions solely by the triplet $(\mathcal{F}, \mathcal{V}, \mathcal{B})$. From these the associated joint density can always be constructed.

Definition 5.17 *(Regular vine copula)* A *regular vine copula* is a regular vine distribution, where all margins are uniformly distributed on [0, 1].

Remark 5.18 (Existence of regular vine distributions when non-symmetric pair copulas are used) The existence result of Theorem 5.15 remains valid for ordered pairs in trees with directed edges. This changes Eq. (5.2) only by specifying which element of an edge $\{a, b\}$ is the head "a" and which is the tail "b" of the directed edge $a \to b$. This was already shown by Bedford and Cooke (2002) without explicitly mentioning it. In particular Theorem 3 of Bedford and Cooke (2002) is valid for this case.

Remark 5.19 (Ordering of pair copula indices) In general we make the following convention: Copula $C_{i,j;i_1,\ldots,i_r}$ has first argument $F_{i|i_1,\ldots,i_k}$ and second argument $F_{j|i_1,\ldots,i_k}$, for $i \neq j$ and additional distinct indices $i_1,\ldots,i_k$ for the conditioning set. Further we arrange the indices both for the conditioned and conditioning set in increasing order.

Ex 5.9 (Example of an R vine distribution in six dimensions) The R-vine density corresponding to R-vine tree structure given in Fig. 5.3 is

$$f_{123456}(x_1,\ldots,x_6) = c_{26;1345} \cdot c_{25;134} \cdot c_{46;135}$$
$$\cdot\ c_{45;13} \cdot c_{24;13} \cdot c_{36;15} \cdot c_{35;1} \cdot c_{14;3} \tag{5.3}$$
$$\cdot\ c_{23;1} \cdot c_{16;5} \cdot c_{15} \cdot c_{34} \cdot c_{13} \cdot c_{12} \cdot c_{56} \cdot f_6 \cdot f_5 \cdot f_4 \cdot f_3 \cdot f_2 \cdot f_1,$$

where we used the abbreviation introduced in (4.17) for the pair copula terms, $f_{1\cdots j}$ for $f_{1\cdots j}(x_1,\cdots,x_j)$ in case of the joint density and f_j for $f_j(x_j)$ in case of the marginal densities. We also assume that the simplifying assumption of Remark 5.12 holds.

Following Remark 5.19 we arrange the indices of the conditioned set and the conditioning set in increasing order. In this example we use

$$c_{36;15} = c_{36;15}(F_{3|15}(x_3|x_1, x_5), F_{6|15}(x_6|x_1, x_5)),$$

which is the same as

$$c_{63;15} = c_{63;15}(F_{6|15}(x_6|x_1, x_5), F_{3|15}(x_3|x_1, x_5)).$$

The joint density (5.3) can also be built by the following recursive decomposition

$$f_{12345} = f_1 \cdot f_{2|1} \cdot f_{3|12} \cdot f_{4|123} \cdot f_{5|1234} \cdot f_{6|12345}.$$

and these conditional densities can be expressed as

$$f_{2|1} = c_{12} \cdot f_2 \tag{5.4a}$$

$$\begin{aligned} f_{3|12} &= c_{23;1} \cdot f_{3|1} \\ &= c_{23;1} \cdot c_{13} \cdot f_3, \end{aligned} \tag{5.4b}$$

$$\begin{aligned} f_{4|123} &= c_{24;13} \cdot f_{4|13} = c_{24;13} \cdot c_{14;3} \cdot f_{4|3} \\ &= c_{24;13} \cdot c_{14;3} \cdot c_{34} \cdot f_4, \end{aligned} \tag{5.4c}$$

$$\begin{aligned} f_{5|1234} &= c_{25;134} \cdot f_{5|134} \\ &= c_{25;134} \cdot c_{45;13} \cdot f_{5|13} \\ &= c_{25;134} \cdot c_{45;13} \cdot c_{35;1} \cdot f_5 \\ &= c_{25;134} \cdot c_{45;13} \cdot c_{35;1} \cdot c_{15} \cdot f_{5|1} \end{aligned} \tag{5.4d}$$

$$\begin{aligned} f_{6|12345} &= c_{26;1345} \cdot f_{6|1345} \\ &= c_{26;1345} \cdot c_{46;135} \cdot f_{6|135} \\ &= c_{26;1345} \cdot c_{46;135} \cdot c_{36;15} \cdot f_{6|15} \\ &= c_{26;1345} \cdot c_{46;135} \cdot c_{36;15} \cdot c_{16;5} \cdot f_{6|5} \\ &= c_{26;1345} \cdot c_{46;135} \cdot c_{36;15} \cdot c_{16;5} \cdot c_{56} \cdot f_6. \end{aligned} \tag{5.4e}$$

Note that we used only pair copulas associated with the edges specified in vine tree sequence given in Fig. 5.3.

To completely specify the R-vine density (5.3), we require the conditional distribution functions, which occur as arguments in the pair copula terms of representation (5.3) For this we can apply recursively Theorem 4.10 and illustrate this for Example 5.9.

Ex 5.10 (*Conditional distribution functions for R-vine in Example 5.9*)   To obtain expressions for the arguments of the bivariate densities, which only require bivariate copulas already used in the construction and thus do not require integration, we have to make a "clever choice" in each step. Figure 5.5 illustrates these choices for Example 5.9.

We give now an example for a choice, which would require further bivariate copula terms than given in the representation (5.3). If we had chosen the copula density $c_{56;13}$ instead of $c_{36;15}$ in (5.4e) the density of neither copulas $C_{16;3}$ nor $C_{36;1}$, from which the argument $F_{6|13}$ can be calculated, would have been included in the repesentation.

But how to systematically make a "clever choice" in each step? The arguments of copulas conditioned on d variables will always have to be expressed by copulas conditioned on $d - 1, \ldots, 1$ variables. It is thus natural to use a bottom-up approach, since possible choices on the "higher" trees depend on the choices made on the "lower" trees. To illustrate the bottom up approach we study the vine tree structure specified in Fig. 5.3.

In the first tree, i.e. T_1, we have all the bivariate copulas of decomposition (5.3), each $C_{i,j}$ represented as an edge linking the univariate margins i and j (shown as nodes). For T_2 we have the bivariate copula indices of the edges in tree T_1 as nodes. The edges in T_2 are the indices of the bivariate copulas conditioned on one variable. They link the nodes from which their arguments are obtained using a recursion based on Theorem 4.10. Continuing in this way all required bivariate copulas in (5.3) can be identified by the five trees. Consider for example the copula density

$$c_{36;15}(F_{3|15}(x_3|x_1, x_5), F_{6|15}(x_6|x_1, x_5)),$$

with arguments obtainable from

$$F_{3|15}(x_3|x_1, x_5) = \frac{\partial C_{35;1}(F_{3|1}(x_5|x_1), F_{5|1}(x_5|x_1))}{\partial F_{5|1}(x_5|x_1)}$$

and

$$F_{6|15}(x_6|x_1, x_5) = \frac{\partial C_{16;5}(F_{1|5}(x_1|x_5), F_{6|5}(x_6|x_5))}{\partial F_{1|5}(x_1|x_5)}.$$

It is represented as edge 36; 15 between nodes 35; 1 and 16; 5.

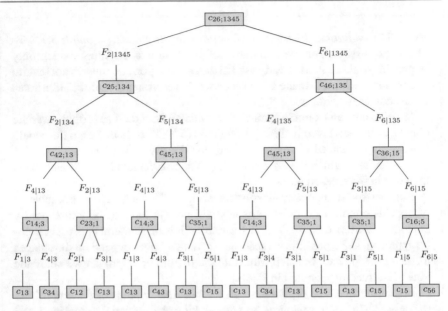

Fig. 5.5 Conditional distribution functions: Following this chart, the arguments of all copulas in decomposition (5.3) for Example 5.9 can be obtained recursively. The pair copulas marked with rectangular shapes are present in the vine decomposition (5.4a–e)

5.4 Simplified Regular Vine Classes

Using the decomposition (4.16) of a joint d dimensional density together with a recursive application of Lemma 1.15 we can represent any multivariate density by pair copulas applied to marginal and conditional distribution functions together with their marginal densities. This general factorization can be considered in a regular vine distribution, if the indices corresponding to the pair copulas used form a regular vine tree sequence. However even in this case the pair copulas will often depend on the values of the conditioning variables. Such regular vine distributions will be called *non simplified*.

In contrast regular vine distributions in which no pair copula depends on the conditioning value are called *simplified* as already mentioned in Remark 5.12. However even in the case of simplified regular vine distributions the conditioning values have an effect, since the arguments given by the conditional distribution functions do depend on the values of the conditioning variables.

It is now interesting to investigate the magnitude of the class of simplified regular vines and copulas. For this it is enough to consider the case of simplified regular vine copulas, since with Sklar's Theorem 1.9 we can easily transform from copulas to distributions. There are several multivariate copulas, which can be represented as a simplified regular vine distribution.

Theorem 5.20 (Multivariate copulas represented as simplified regular vine copulas) *The following multivariate copula classes can be represented as simplified regular vine copulas:*

1. *The class of the multivariate Gaussian copulas. Here, the pair copulas are bivariate Gaussian copulas with dependence parameter given by the corresponding partial correlation.*
2. *The class of the multivariate Student t copulas with v degrees of freedom and dependence matrix R a correlation matrix. Here the pair copulas of tree T_j are bivariate Student t copulas with $v + j - 1$ degrees of freedom and dependence parameter the corresponding partial correlation.*
3. *The class of multivariate Clayton copulas.*

Proof To proof the first statement, we note that the Gaussian distribution is closed under conditioning (see Example 1.4). In particular, only the conditional mean depends on the conditioning variable, which implies that the copula associated with this conditional distribution is independent of the conditioning value (cf. Lemma 1.13). The parameters of these copulas are given by conditional correlations, which are equal to partial correlations in the class of elliptical distributions.

For the second statement we also use the closure under conditioning for the multivariate t distribution, however here also the covariance of the t distribution depends on the conditioning value (see Example 1.7). However this dependence is only through a scaling factor. Therefore the overall dependence on the conditioning values is only through location and scale. Since copulas are invariant under monotone transformations (cf. Lemma 1.13) they are also invariant under location and scale transformations, thus the copula corresponding to the conditional distribution does not depend on the conditioning values. The dependence parameters of the corresponding pair copula are conditional correlations and thus partial correlations. The degrees of freedom depend on the number of conditioning values are discussed in Example 1.7.

The proof of the third statement uses the fact that the multivariate Clayton copula is the copula of the multivariate Burr distribution. The multivariate Burr distribution is also closed under conditioning (Takahashi 1965). For more details see Exercise 1.4.

In general, we can say that the class of simplified regular vine distributions is very flexible, since the number of vine tree structures is huge (see Remark 5.8). To illustrate this, the precise numbers are given for different dimensions in Table 5.2.

Further, we can use any bivariate copula family (parametric or nonparametric) as pair copulas. Additionally Nagler and Czado (2016) have shown, that the use of nonparametric pair copulas in simplified regular vines avoids the curse of dimensionality in multivariate density estimation. Corresponding kernel density estimation

Table 5.2 Regular vine tree sequences: Number of different regular vine tree sequences for different dimensions

d	Number of vine tree structures
3	3
4	24
5	480
6	23040
7	2580480
8	660602880
9	380507258880
10	487049291366400
15	19761016339043027619853253581209600

methods have been implemented in the R library kdecopula of Nagler (2017a) and a general comparison of non parametric pair copula estimation in the context of simplified vines is available in Nagler et al. (2017).

See also Stöber et al. (2013) for the effects of approximating non simplified distributions by simplified regular vine distributions and Acar et al. (2012) for local polynomial smoothing approach to fit non simplified vine distributions. However their approach is at the moment restricted to three dimensions and involves multidimensional smoothing, which might be difficult to extend to higher dimensions. A more feasible approach in higher dimensions using a generalized additive model specification has been developed by Vatter and Nagler (2018). Finally recall that many shapes for bivariate margins are possible in simplified vines as shown in Fig. 4.1.

For Gaussian regular vines there is a very useful relationship between partial correlations and ordinary correlations.

Remark 5.21 (Relationship between partial correlations and correlation matrix) The correlations contained in a correlation matrix have to be chosen such, that the resulting matrix is positive definite. In contrast, we can choose partial correlations arbitrarily in $[-1, 1]$ in a regular Gaussian vine distribution and there exists according to Theorem 4.4 of Bedford and Cooke (2002) a one-to-one relationship between the space of partial correlations of regular Gaussian vines and the space of positive definite correlation matrices. For an example of this relationship see Exercise 2.7. Joe (2006) applies this relationship to use regular Gaussian vines to generate uniformly on the space of correlation matrices.

For the determinant of the correlation matrix of a Gaussian distribution Kurowicka and Cooke (2006) showed the following relationship:

Theorem 5.22 (Relationship between the determinant of a correlation matrix and the partial correlations of a Gaussian vine distribution) *Let $\mathcal{D}$ be the determinant of the correlation matrix of multivariate Gaussian distribution in d dimensions with $\mathcal{D} > 0$. Then for any Gaussian vine distribution with partial correlations $\rho_{\mathcal{C}_{e,a},\mathcal{C}_{e,b};D_e}$ associated to the edge $e = (\mathcal{C}_{e,a}, \mathcal{C}_{e,b}; D_e) \in V$ the following relationship holds*

$$\mathcal{D} = \prod_{i=1}^{d-1} \prod_{e \in E_i} \rho_{\mathcal{C}_{e,a},\mathcal{C}_{e,b};D_e}.$$

Proof The proof can be found on p. 126 of Kurowicka and Cooke (2006).   □

5.5   Representing Regular Vines Using Regular Vine Matrices

To develop inference methods for arbitrary R-vines we need a way to store the vine tree sequence in the computer. For this we use matrices first introduced by Kurowicka (2009) and extended for likelihood calculations by Dißmann (2010), Dißmann et al. (2013).

To store the vine tree structure $\{\mathcal{C}_{e,a}, \mathcal{C}_{e,b}; D_e, e \in T_j, j = 1, ..., d\}$ the associated indices are stored in an upper triangular matrix. Upper triangular matrices are utilized, since they allow, that sums in the log likelihood of R-vine distributions can be represented by increasing indices in contrast, when lower triangular matrices are used. This has already been followed in Chap. 5 of Stöber and Czado (2017), while the work by Dißmann et al. (2013), Dißmann (2010) uses lower triangular matrices. The R library `VineCopula` of Schepsmeier et al. (2018) can handle both lower and upper triangular matrices.

Definition 5.23 (*Regular vine matrix*) Let M be an upper triangular matrix with entries $m_{i,j}$ for $i \leq j$. The elements $m_{i,j}$ can have values between 1 to d. A matrix M is called a *regular vine matrix*, if it satisfies the following conditions:

1. $\{m_{1,i}, \ldots, m_{i,i}\} \subset \{m_{1,j}, \ldots, m_{j,j}\}$ for $1 \leq i < j \leq d$ (The entries of a specific column are also contained in all columns right of this column.)
2. $m_{i,i} \notin \{m_{1,i-1}, \ldots, m_{i-1,i-1}\}$ (The diagonal entry of a column does not appear in any column further to the left.)
3. For $i = 3, \ldots, d$ and $k = 1, \ldots, i - 1$ there exist (j, ℓ) with $j < i$ and $\ell < j$ such that

$$\left\{m_{k,i}, \{m_{1,i}, \ldots, m_{k-1,i}\}\right\} = \left\{m_{j,j}, \{m_{1,j}, \ldots, m_{\ell,j}\}\right\} \quad \textbf{or}$$

$$\left\{m_{k,i}, \{m_{1,i}, \ldots, m_{k-1,i}\}\right\} = \left\{m_{\ell,j}, \{m_{1,j}, \ldots, m_{\ell-1,j}, m_{j,j}\}\right\}.$$

The last assumption in this definition is the analogue of the proximity condition for regular vine trees. With this definition of an R-vine matrix it was shown that there is a bijection between regular vine trees and regular vine matrices. Corresponding algorithms are given below and are taken from Stöber and Czado (2017). Before we present these we illustrate the algorithms in examples.

Ex 5.11 (Construction of a regular vine matrix for the vine tree sequence of Example 5.9) In the first step we select one of the entries of the conditioned set of the single edge in tree T_{d-1}, i.e. 2 or 6 from edge 2, 6; 1, 3, 4, 5 of tree T_5 in Example 5.3, and put it in the lower right corner of a d dimensional matrix.

Selecting for example the element 6 we write down all indices, which are in the conditioned sets of an edge together with 6 (bolded in Fig. 5.6). These are the numbers 2, 4, 3, 1, 5. We order them in this way, since 2 occurs in T_5, 4 in T_4, 3 in T_3, 1 in T_2 and 5 in T_1. Choosing 6 together with such a number and the numbers above this entry identifies a particular edge in the vine tree sequence. For example the entries 6 and 4 and the entries 3, 1, 5 above in the last column identifies the edge 4, 6; 1, 3, 5. In summary the edge 2, 6; 1, 3, 4, 5, the edge 4, 6; 1, 3, 5, the edge 3, 6; 1, 5, the edge 1, 6; 5 and the edge 5, 6 are identified by the last column. Recall that we order the entries of the conditioned and conditioning set in increasing order. More generally the entries $m_{6,6}$ and $m_{j,6}$ identify the edge $m_{6,6}, m_{j,6}; m_{1,6}, \ldots, m_{j-1,6}$ for $j = 2, \ldots, d-1$ up to ordering of the indices.

Therefore all pair copula terms needed to characterize the dependence of X_6 on $X_1, \ldots, X_5$ is stored in the last column of the matrix M. We remove now all nodes and edges of the vine tree sequence containing the index 6. These are exactly the ones which we have just identified for 6 in the conditioned set, and we end up with the following reduced vine tree sequence given in Fig. 5.7.

With this second vine tree sequence we redo the above procedure, selecting for example 5 in tree T_4 of Fig. 5.7 and selecting it as diagonal element of the second last column of the matrix. Adding the entries, which are in the conditioned sets together with 5 ordered by the tree level they are occurring in, the matrix is filled as shown on the left panel of Fig. 5.7.

Continuing, the selected nodes in the second last column are removed and the resulting reduced vine tree sequence is shown in Fig. 5.8. These steps are repeated until all nodes of the original vine tree sequence have been removed.

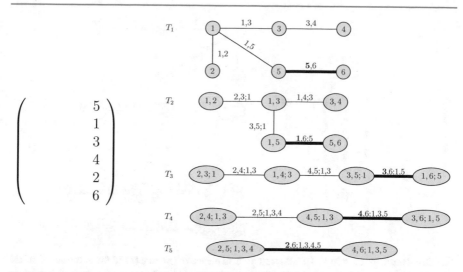

Fig. 5.6 R-vine matrix: Construction of the last column for the R-vine matrix corresponding to Example 5.9. We select 6 in edge 2, 6; 1, 3, 4, 5, and all entries which are in a conditioned set together with 6

This gives the final vine regular matrix as

$$
\begin{pmatrix}
4 & 4 & 3 & 1 & 1 & 5 \\
  & 3 & 4 & 3 & 3 & 1 \\
  &   & 1 & 4 & 4 & 3 \\
  &   &   & 2 & 2 & 4 \\
  &   &   &   & 5 & 2 \\
  &   &   &   &   & 6
\end{pmatrix}. \tag{5.5}
$$

Following these steps we can always determine an R-vine matrix once a regular vine tree sequence is given. This allows the conditioned and conditioning sets of the edges to be stored in the following way: The diagonal entry $m_{k,k}$ of each row k is the first entry of all conditioning sets of the entries, which have been deleted from the vine filling up the row. The entries above the diagonal corresponds to an edge e_i with conditioned set $\mathcal{C}_{e_i} = \{m_{k,k}, m_{i,k}\}$, $i < k$. The proximity condition implies that the associated conditioning set of e_i is $D_{e_i} = \{m_{i-1,k}, \ldots, m_{1,k}\}$. A general algorithm for computing the associated regular matrix was given in Stöber and Czado (2017) and is reprinted in Algorithm 5.1.

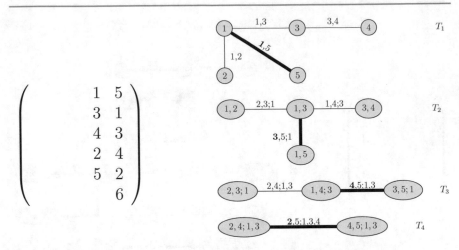

$$\begin{pmatrix} 1 & 5 \\ 3 & 1 \\ 4 & 3 \\ 2 & 4 \\ 5 & 2 \\ & 6 \end{pmatrix}$$

Fig. 5.7 R-vine matrix: The reduced vine sequence (right panel) and the construction of the second last column (left panel) after the first step of constructing the associated regular vine matrix

Fig. 5.8 R-vine matrix: Construction of the third column in the R-vine matrix: Here we select 2 in the edge 2, 4; 1, 3 and then all entries which are in a conditioned set together with 2

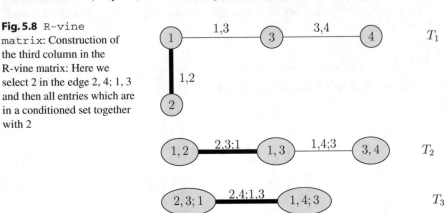

Remark 5.24 (Non uniqueness of the R-vine matrix) Since at each step you can choose between the two elements of the conditioned set, the resulting R-vine matrix is not unique. However it encodes all edges in a vine tree sequence and thus is highly useful for statistical programming with regular vines.

We now consider the reverse problem of drawing the associated R-vine tree sequence from a given R-vine matrix. This algorithm inverts the steps of Algorithm 5.1 by adding for each identified edge from the regular vine matrix the associated nodes and edges and is given as Algorithm 5.2, which is taken from Stöber and Czado (2017).

Algorithm 5.1 (*Computing a regular vine matrix for a regular vine tree sequence* $\mathcal{V}$) The input of the algorithm is a regular vine tree sequence $\mathcal{V} = (T_1, \ldots, T_{d-1})$ and the output will be a regular vine matrix M.

$$\mathcal{X} := \{\} \tag{1}$$

$\quad$ FOR $i = d, \ldots, 3$ $\hfill (2)$

$\qquad$ Choose $x, \tilde{x}, D$ with $\tilde{x} \notin \mathcal{X}$ and $|D| = i - 2$ such that there is an edge e

$\qquad$ with $C_e = \{x, \tilde{x}\}, D_e = D$ $\hfill (3a)$

$\qquad m_{i,i} := x, m_{i-1,i} := \tilde{x}$ $\hfill (3b)$

$\qquad$ FOR $k = i - 2, \ldots, 1$ $\hfill (4)$

$\qquad\quad$ Choose $\hat{x}$ such that there is an edge e with $C_e = \{x, \hat{x}\}$ and

$\qquad\quad |D_e| = k - 1$ $\hfill (5a)$

$\qquad\quad m_{k,i} := \hat{x}$ $\hfill (5b)$

$\qquad$ END FOR

$\qquad \mathcal{X} := \mathcal{X} \cup \{x\}$ $\hfill (6)$

$\quad$ END FOR

$\quad$ Choose $x, \tilde{x} \in \{1, \ldots, d\} \setminus \mathcal{X}$ $\hfill (7a)$

$\quad m_{2,2} := x, m_{1,2} := \tilde{x}, m_{1,1} := \tilde{x}$ $\hfill (7b)$

$\quad$ RETURN $M := (m_{k,i} | k = 1, \ldots d, \ k \leq i)$ $\hfill (8)$

(Taken from Stöber and Czado (2017). With permission of ©World Scientific Publishing Co. Pte. Ltd. 2017.)

Algorithm 5.2 (*Construction of a tree sequence from an R-vine matrix* M) The input of the algorithm is a d-dimensional regular vine matrix M and the output will be a regular vine $\mathcal{V} = (T_1, \ldots, T_{d-1})$.

$$N_1 := \{1, \ldots, d\} \tag{1a}$$

$E_2 := \{\}, \ldots, E_{d-1} := \{\}$ $\hfill (1b)$

$E_1 := \{m_{2,2}, m_{1,2}\}$ $\hfill (1c)$

$\quad$ FOR $i = 3, \ldots, d$ $\hfill (2)$

$\qquad e_1^i := \{m_{i,i}, m_{1,i}\}$ $\hfill (3a)$

$\qquad E_1 := E_1 \cup \{e_1^i\}$ $\hfill (3b)$

$\qquad$ FOR $k = 1, \ldots, i - 2$ $\hfill (4)$

$\qquad\quad$ Select $a_k \in E_k$ with $A_{a_k} = \{m_{1,i}, \ldots, m_{1+k,i}\}$ $\hfill (5a)$

$\qquad\quad e_{k+1}^i := \{e_k^i, a_k\}$ $\hfill (5b)$

$\qquad\quad E_{k+1} := E_{k+1} \cup \{e_{k+1}^i\}$ $\hfill (5c)$

$\qquad$ END FOR

$\quad$ END FOR

$\quad$ RETURN $\mathcal{V} := \left(T_1 := (N_1, E_1), T_2 := (E_1, E_2), \ldots, T_{d-1} := (E_{d-2}, E_{d-1}) \right)$ $\hfill (6)$

(Taken from Stöber and Czado (2017). With permission of ©World Scientific Publishing Co. Pte. Ltd. 2017.)

Col M   Edge   Tree structure

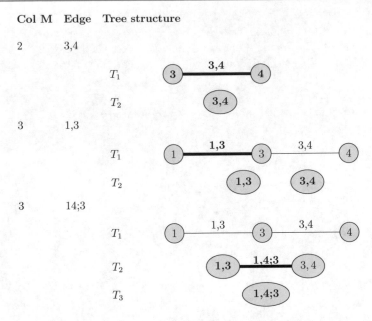

2   3,4

3   1,3

3   14;3

Fig. 5.9 R-vine tree construction: First three steps in building up the associated vine tree sequence for R-vine matrix given in (5.5)

The order in which edges are added in Algorithm 5.2 is chosen such that it coincides with the set notation of Definition 5.4. We illustrate Algorithm 5.2 by the following example.

Ex 5.12 (Constructing a vine tree sequence from the regular vine matrix (5.5))
The algorithm starts in column 2 of the regular vine matrix specified in (5.5), adding an edge between the two entries in this column, in our example 3 and 4 (see Fig. 5.9 first row). Further, node 3, 4 is added to tree T_2. It then moves one column to the right, adding edge 1, 3 in tree T_1, as well as node 1, 3 to tree T_2 (see Fig. 5.9 second row). Then the edge 1, 4; 3 between 1, 3 and 3, 4 in T_2 and the node 1, 4; 3 in tree T_3 are added (see Fig. 5.9 third row). Adding the next three edges identified by column 4 of the R-vine matrix (5.5) are illustrated in Fig. 5.10 These steps are repeated until the whole R-vine tree sequence is rebuilt after row d.

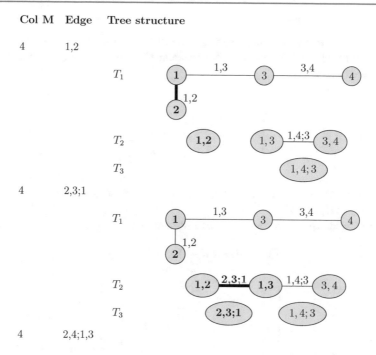

Col M   Edge   Tree structure

4	1,2	
4	2,3;1	
4	2,4;1,3	

Fig. 5.10 R-vine tree construction: Adding the information from the three edges identified from column 4 of the R-vine matrix (5.5) to the associated R-vine tree sequence

Remark 5.25 All algorithms presented in this section work exactly the same way if we replace unordered sets with ordered pairs, i.e. if we switch to directed graphs. Introducing the ordering we get a one-to-one correspondence between R-vine matrices and directed regular vines (since x and $\tilde{x}$ in Algorithm 5.1 become distinguishable). Therefore we can also assign non-symmetric copulas to an R-vine as soon as it is parametrized by an R-vine matrix with the convention that the first argument of the copula always corresponds to the diagonal entry of the R-vine matrix.

With the algorithms outlined above, all necessary tools to describe the structure of a pair copula construction have now been developed. To conclude this section, Example 5.13 shows the corresponding matrices for the special C- and D-vine tree sequences from Definition 5.9.

Ex 5.13 (Regular vine matrices for C-vine and D-vines) After permutation of $(1, \ldots d)$ the R-vine matrix of a d dimensional C-vine tree sequence can always be expressed as

$$
\begin{pmatrix}
1 \cdots & 1 & 1 & 1 \\
\ddots & \vdots & \vdots & \vdots \\
& d-2 & d-2 & d-2 \\
& & d-1 & d-1 \\
& & & d
\end{pmatrix}.
\tag{5.6}
$$

In this case the root node of tree T_1 is 1, the root node of tree T_2 is $1, 2 \ldots$ of tree T_{d-1} is $d, d-1; 1, \ldots, d-2$.

The R-vine matrix for a D-vine can be written as

$$
\begin{pmatrix}
1 \cdots & d-4 & d-3 & d-2 & d-1 \\
\ddots & \vdots & \vdots & \vdots & \vdots \\
& d-3 & 1 & 2 & 3 \\
& & d-2 & 1 & 2 \\
& & & d-1 & 1 \\
& & & & d
\end{pmatrix}.
\tag{5.7}
$$

As mentioned in Remark 5.10 the ordering of the node $1, \ldots, d$ in tree T_1 determines all further trees of the vine tree sequence.

Ex 5.14 (Interpreting an regular vine matrix given as lower triangular matrix) Suppose the regular vine matrix in lower triangular matrix notation is given by

$$
\begin{pmatrix}
2 \\
5 & 3 \\
3 & 5 & 4 \\
1 & 1 & 5 & 5 \\
4 & 4 & 1 & 1 & 1
\end{pmatrix}
\tag{5.8}
$$

For example the second column of matrix (5.8) identifies the edge 3, 5; 1, 4, the edge 3, 1; 4 and the edge 3, 4. Thus the following edges of the R-vine tree sequence are identified

- Tree T_1: $\{2, 4\}, \{3, 4\}, \{4, 1\}, \{5, 1\}$
- Tree T_2: $\{2, 1; 4\}, \{3, 1; 4\}, \{4, 5; 1\}$

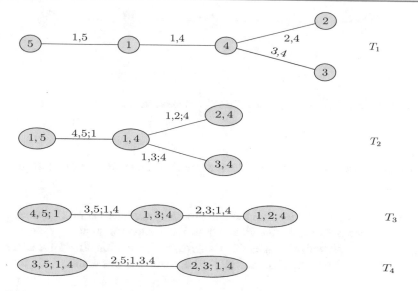

Fig. 5.11 R-vine matrix: Regular vine sequence corresponding to lower triangular regular vine matrix (5.8)

- Tree T_3: $\{2, 3; 1, 4\}$, $\{3, 5; 1, 4\}$
- Tree T_4: $\{2, 5; 3, 1, 4\}$.

The corresponding R-vine tree sequence is given in Fig. 5.11.

5.6   Exercises

Exer 5.1

R-vine tree sequence plots in VineCopula: Use the function plot for an RVM object in the R library VineCopula to plot the R vine tree sequence given in Fig. 5.3 using the R vine matrix specified in (5.5).

Exer 5.2

Construction of a lower triangular regular vine matrix for C- and D-vine tree sequences: Derive the lower triangular regular vine matrices corresponding to a C-vine tree sequence in five dimensions. Do the same for a D-vine tree sequence in five dimensions.

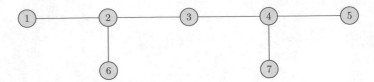

Fig. 5.12 R-vine tree sequence: Tree T_1 considered in Exercise 5.3

Exer 5.3
Proximity in vine structures: Consider the tree T_1 of a regular vine structure given in Fig. 5.12.

(i) Draw a graph with all edges that the proximity condition allows for tree T_2.
(ii) Select one permissible tree T_2 and draw it in a separate graph.
(iii) Draw a graph with all edges that the proximity condition allows for a tree T_3, given the selected tree T_2. Add all the edge labels to this graph.

Exer 5.4
Construction of a regular vine tree sequence from a R-vine matrix: Draw the regular vine structure that corresponds to the regular vine matrix

$$M = \begin{pmatrix} 7 & & & & & & \\ 3 & 6 & & & & & \\ 4 & 3 & 5 & & & & \\ 5 & 4 & 3 & 4 & & & \\ 6 & 5 & 4 & 3 & 3 & & \\ 2 & 2 & 2 & 2 & 2 & 2 & \\ 1 & 1 & 1 & 1 & 1 & 1 & 1 \end{pmatrix}.$$

Is this not only a regular vine but also a C-vine or a D-vine? Justify your answer. Additionally, determine how many different regular vine matrices exist that correspond to this regular vine.

Exer 5.5
Construction of a regular vine density from a regular vine tree sequence: Provide the formula for the density $f(x_1, \ldots, x_7)$ of the regular vine distribution with the underlying regular vine tree sequence specified in Fig. 5.13 assuming that all pair copulas in tree T_3 until tree T_5 are independence copulas.

Exer 5.6
Conditional distributions in vine copula: Express the conditional distribution U_7 given U_3, U_4, U_5, i.e. $C_{7|345}(u_7|u_3, u_4, u_5)$ in terms of h-functions of pair-copulas specified by a vine copula with vine tree sequence given in Fig. 5.13.

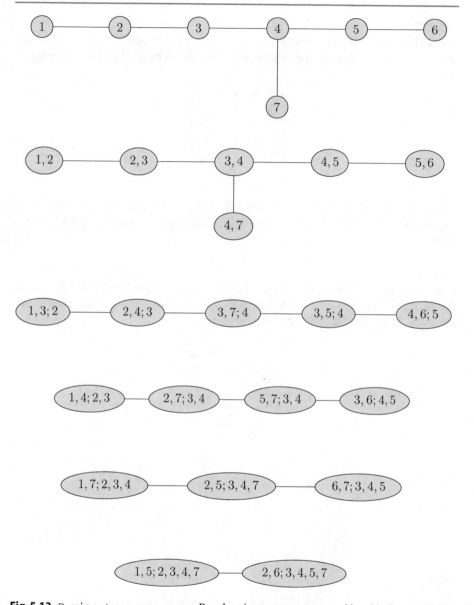

Fig. 5.13 R-vine tree sequence: Regular vine tree sequence considered in Exercise 5.5

Exer 5.7

Derive R-vine tree sequence and R-vine density from a specified R-vine matrix: Derive
the corresponding R-vine tree sequence and R-vine density from the following R-
vine matrix

$$\begin{pmatrix} 2 & 0 & 0 & 0 & 0 \\ 5 & 3 & 0 & 0 & 0 \\ 3 & 5 & 4 & 0 & 0 \\ 1 & 1 & 5 & 5 & 0 \\ 4 & 4 & 1 & 1 & 1 \end{pmatrix}.$$

Exer 5.8

Construction of an R vine matrix from a R vine tree sequence: Consider again the
regular vine tree sequence specified in Fig. 5.13. Derive a corresponding regular vine
matrix.

Exer 5.9

Derivation of conditional distribution functions in Example 5.9: Derive the conditional distribution functions in (5.4) of Example 5.9 associated with the conditional
densities on the left hand side of the equation.

Simulating Regular Vine Copulas and Distributions

6

6.1 Simulating Observations from Multivariate Distributions

For simulation from a d-dimensional distribution function $F_{1,\ldots,d}$ with conditional distribution functions $F_{j|1,\ldots,j-1}(\cdot|x_1,\ldots,x_{j-1})$ and their inverses $F_{j|1,\ldots,j-1}^{-1}$ $(\cdot|x_1,\ldots,x_{j-1})$ for $j=2,\ldots,d$ we can use iterative inverse probability transformations. In particular, a multivariate transformation introduced by Rosenblatt (1952) and studied more generally by Rüschendorf (1981) is utilized and stepwise inverted. It is also called the *Rosenblatt transform*. More specifically Theorem 6.1 holds.

Theorem 6.1 (Simulating from a d variate distribution) *To obtain a sample* $x_1,\ldots,x_d$ *from* $F_{1,\ldots,d}$, *the following steps can be performed:*

$$\textbf{\textit{First}}: Sample \ w_j \overset{i.i.d.}{\sim} U[0;1], \quad j=1,\ldots,d$$

$$\textbf{\textit{Then}}: x_1 := F_1^{-1}(w_1)$$

$$x_2 := F_{2|1}^{-1}(w_2|x_1)$$

$$\vdots$$

$$x_d := F_{d|d-1,\ldots,1}^{-1}(w_d|x_{d-1},\ldots,x_1).$$

Then, $(x_1,\ldots,x_d)$ *is a random sample of size* $n=1$ *from the distribution* $F_{1,\ldots,d}$.

Proof First, x_1 is a realization from the marginal distribution of X_1 and x_2 is a realization from the conditional distribution of X_2 given $X_1 = x_1$ by applying the

© Springer Nature Switzerland AG 2019
C. Czado, *Analyzing Dependent Data with Vine Copulas*, Lecture Notes in Statistics 222, https://doi.org/10.1007/978-3-030-13785-4_6

probability integral transformation. Therefore, it follows that (x_1, x_2) is a sample from the bivariate marginal distribution (X_1, X_2). Induction can be used to proof the remainder of the theorem.

This shows that it is necessary to determine conditional distribution functions and to be able to sample from them. We will now apply the transformation in Theorem 6.1 to sample from regular vine copulas.

Simulations from regular vine distributions have first been discussed in Bedford and Cooke (2001), Bedford and Cooke (2002) but without giving programmable algorithms. Aas et al. (2009), Kurowicka and Cooke (2005) developed sampling algorithms for C-vines and D-vines. While Kurowicka and Cooke (2005) also gave hints on how to treat the general R-vine case, it was Dißmann (2010), who demonstrated how to write a sampling algorithm for the general R-vine using the matrix notation from Sect. 5.5. The algorithms we present are slightly corrected versions from Stöber and Czado (2017). These are based on Dißmann et al. (2013), where some redundant calculations have been omitted.

6.2 Simulating from Pair Copula Constructions

Adapting Theorem 6.1 to copulas gives the following procedure:

Theorem 6.2 (Simulating from a multivariate copula) *Perform the following steps to obtain a sample* $u_1, \ldots, u_d$ *from a d variate copula:*

$$\textbf{\textit{First}}: \textit{Sample } w_j \overset{i.i.d.}{\sim} U[0; 1], \quad j = 1, \ldots, d$$
$$\textbf{\textit{Then}}: u_1 := w_1$$
$$u_2 := C_{2|1}^{-1}(w_2|u_1)$$
$$\vdots$$
$$u_d := C_{d|d-1,\ldots,1}^{-1}(w_d|u_{d-1}, \ldots, u_1)$$

To determine the conditional distribution functions $C_{j|j-1,\ldots,1}$, $j = 1, \ldots, d$ needed for pair copula constructions, we will use the recursive relationship for the conditional distribution function given in Theorem 4.10 together with the h-function introduced in Definition 1.16. This gives an iterative expression using h-functions for the desired conditional distribution function, which can easily inverted recursively. More precisely, we use the notation $h_{1|2}(u_1|u_2; \theta_{12})$ for a $h_{1|2}$-function with parameters θ_{12} of a specified bivariate copula C_{12}. The general notation utilized in this chapter is now given.

Definition 6.3 (*General notation of h-functions of bivariate parametric copulas*) For the bivariate copula $C_{ij}(u_i, u_j; \theta_{ij})$ with parameter θ_{ij}, we define the *h*-functions

$$h_{i|j}(u_i|u_j; \theta_{ij}) := \frac{\partial}{\partial u_j} C_{ij}(u_i, u_j; \theta_{ij}) \tag{6.1}$$

$$h_{j|i}(u_j|u_i; \theta_{ij}) := \frac{\partial}{\partial u_i} C_{ij}(u_i, u_j; \theta_{ij}). \tag{6.2}$$

For the parametric pair copula $C_{e_a,e_b;D_e}(w_1, w_2; \theta_{e_a,e_b;D_e})$ in a simplified regular vine corresponding to the edge $e_a, e_b; D_e$, we introduce the notation

$$h_{e_a|e_b;D_e}(w_1|w_2; \theta_{e_a,e_b;D_e}) := \frac{\partial}{\partial w_2} C_{e_a,e_b,D_e}(w_1, w_2; \theta_{e_a,e_b;D_e}) \tag{6.3}$$

$$h_{e_b|e_a;D_e}(w_2|w_1; \theta_{e_a,e_b;D_e}) := \frac{\partial}{\partial w_1} C_{e_a,e_b;D_e}(w_1, w_2; \theta_{e_a,e_b;D_e}). \tag{6.4}$$

We need the two versions of the *h*-functions since the bivariate copula indices are usually ordered such as C_{13} and not as C_{31}. Thus to determine the conditional distribution functions $C_{1|3}$ and $C_{3|1}$ both versions are required.

Ex 6.1 (*h-functions associated with pair copulas*) For the edges 23; 1 and 24; 13 of the R-vine considered in Example 5.9 with pair copulas $C_{23;1}$ and $C_{24;13}$, respectively, we can express associated *h*-functions as follows:

$$h_{2|3;1}(w_1|w_2; \theta_{23;1}) := \frac{\partial}{\partial w_2} C_{23;1}(w_1, w_2; \theta_{23;1})$$

$$h_{2|4;13}(w_1|w_2; \theta_{24;13}) := \frac{\partial}{\partial w_2} C_{24;13}(w_1, w_2; \theta_{24;13}).$$

Before deriving simulation algorithms for arbitrary dimensions, we first consider a three-dimensional example.

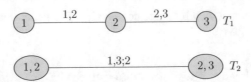

Fig. 6.1 R-vine tree sequence for $d = 3$: The only possible three-dimensional regular vine tree sequence (up to permutations of the nodes)

Ex 6.2 (Simulating from a three-dimensional pair copula construction) In three dimensions, there is up to permutation of the nodes in tree T_1 only one possible regular vine structure, displayed in Fig. 6.1. Having independently drawn w_1, w_2, and w_3 from the uniform distribution, we directly set $u_1 := w_1$. For u_2 we have

$$C_{2|1}(u_2|u_1) = \frac{\partial C_{12}(u_1, u_2)}{\partial u_1} = h_{2|1}(u_2|u_1; \theta_{12}).$$

Thus, we set $u_2 := h_{2|1}^{-1}(w_2|u_1; \theta_{12})$. The calculation for u_3 is a bit more involved. The conditional density $c_{3|12}(u_3|u_1, u_2)$ of U_3 given $U_2 = u_2$ and $U_1 = u_1$ can be expressed as

$$c_{3|12}(u_3|u_1, u_2) = c_{13;2}(C_{1|2}(u_1|u_2), C_{3|2}(u_3|u_2))c_{3|2}(u_3|u_2).$$

Therefore the corresponding distribution function can be derived by integration applying (6.1), (6.2) and (6.4) as

$$
\begin{aligned}
C_{3|12}(u_3|u_1, u_2) &= \int_0^{u_3} \frac{\partial^2}{\partial w_1 \partial w_2} C_{13;2}(w_1, w_2)|_{w_1 = C_{1|2}(u_1|u_2), w_2 = C_{3|2}(v_3|u_2)} \frac{\partial w_2}{\partial v_3} dv_3 \\
&= \frac{\partial}{\partial w_1} C_{13;2}(w_1, C_{3|2}(u_3|u_2))|_{w_1 = C_{1|2}(u_1|u_2)} \\
&= h_{3|1;2}\left(C_{3|2}(u_3|u_2)|C_{1|2}(u_1|u_2); \theta_{13;2}\right) \\
&= h_{3|1;2}\left(h_{3|2}(u_3|u_2; \theta_{23})|h_{1|2}(u_1|u_2; \theta_{12}); \theta_{13;2}\right).
\end{aligned}
$$

This implies

$$C_{3|12}^{-1}(w_3|u_1, u_2) = h_{3|2}^{-1}\left(h_{3|1;2}^{-1}\left(w_3|h_{1|2}(u_1|u_2, \theta_{12}), \theta_{13;2}\right)|u_2, \theta_{23}\right).$$

To generalize this procedure to arbitrary dimensions, we need to find an algorithmic way of selecting the appropriate arguments for the inverse h-functions in each step. First, we discuss the simulation algorithm for C- and D-vine copulas before considering the general case of regular vine copulas.

6.3   Simulating from C-vine Copulas

Before we discuss *simulating from C-vine copulas* in arbitrary dimensions, we introduce some useful abbreviations. To shorten notation, we use $i : j$ for the vector $(i, \ldots, j)$ of the indices and $\boldsymbol{u}_{i:j}$ for $(u_i, \ldots, u_j)$, respectively.

To further simplify notation, we omit in the following the dependence of the h-function on the copula parameter, i.e., we write $h_{i|j}(u_i|u_j)$ instead of $h_{i|j}(u_i|u_j; \theta_{ij})$. We start our discussion by first considering the four-dimensional case.

Ex 6.3 (*Simulating from a four-dimensional C-vine*) To simulate from a four-dimensional C-vine copula, we need the inverses of the following conditional distributions: $C_{2|1}, C_{3|1:2}$ and $C_{4|1:3}$. For example, we can express the conditional distribution function $C_{4|1:3}$ as the derivative with regard to first argument of the bivariate conditional copula $C_{34;1:2}$ evaluated at the associated univariate conditional distribution functions $C_{3|1:2}$ and $C_{4|1:2}$. A similar case was already discussed in Example 4.6. This allows us to calculate the required conditional distribution functions as follows:

$$
\begin{aligned}
C_{4|1:3}(u_4|\boldsymbol{u}_{1:3}) &= h_{4|3;1:2}(C_{4|1:2}(u_4|\boldsymbol{u}_{1:2})|C_{3|1:2}(u_3|\boldsymbol{u}_{1:2})) \\
&= h_{4|3;1:2}\left(h_{4|2;1}\left(h_{4|1}(u_4|u_1)|h_{2|1}(u_2|u_1)\right)|C_{3|1,2}(u_3|\boldsymbol{u}_{1:2})\right) \\
C_{3|1:2}(u_3|\boldsymbol{u}_{1:2}) &= h_{3|2;1}(C_{3|1}(u_3|u_1)|C_{2|1}(u_2|u_1)) \\
&= h_{3|2;1}(h_{3|1}(u_3|u_1)|h_{2|1}(u_2|u_1)) \\
C_{2|1}(u_2|u_1) &= h_{2|1}(u_2|u_1).
\end{aligned}
$$

Recall that we ignore the parameters associated with the h-functions to make the expressions more compact. Now, we derive the associated inverse functions as

$$
\begin{aligned}
C_{4|1:3}^{-1}(u_4|\boldsymbol{u}_{1:3}) &= h_{4|1}^{-1}\left(h_{4|2;1}^{-1}\left(h_{4|3;1:2}^{-1}\left(u_4|C_{3|1:2}(u_3|\boldsymbol{u}_{1:2})\right)|C_{2|1}(u_2|u_1)\right)|u_1\right) \\
C_{3|1:2}^{-1}(u_3|\boldsymbol{u}_{1,2}) &= h_{3|1}^{-1}\left(h_{3|2;1}^{-1}\left(u_3|h_{2|1}(u_1|u_1)\right)|u_1\right) \\
C_{2|1}^{-1}(u_2|u_1) &= h_{2|1}^{-1}(u_2|u_1).
\end{aligned}
$$

We can now apply Theorem 6.2. In particular, we first simulate four i.i.d. uniformly distributed realizations $w_1, \ldots, w_4$. Then u_1, u_2, u_3, and u_4 is a realization from the C-vine, where

$u_1 := w_1$ (realization from uniform)

$u_2 := h_{2|1}^{-1}(w_2|u_1)$ (realization from $C_{2|1}(\cdot|u_1)$)

$$u_3 := h_{3|1}^{-1}\left(h_{3|2;1}^{-1}\left(w_3\big|h_{2|1}(u_2|u_1)\right)\big|u_1\right)$$
$$= h_{3|1}^{-1}\left(h_{3|2;1}^{-1}\left(w_3\big|h_{2|1}(h_{2|1}^{-1}(w_2|u_1|u_1))\right)\big|u_1\right)$$
$$= h_{3|1}^{-1}\left(h_{3|2;1}^{-1}\left(w_3|w_2\right)\big|w_1\right)$$

(realization from $C_{3|1,2}(\cdot|u_1, u_2)$)

$$u_4 := h_{4|1}^{-1}\left(h_{4|2;1}^{-1}\left(h_{4|3;1:2}^{-1}\left(w_4\big|C_{3|1:2}(u_3|u_{1:2})\right)\big|C_{2|1}(u_2|u_1)\right)\big|u_1\right)$$
$$= h_{4|1}^{-1}\left(h_{4|2;1}^{-1}\left(h_{4|3;1:2}^{-1}\left(w_4|w_3\right)\big|w_2\right)\big|w_1\right).$$

(realization from $C_{4|1,2,3}(\cdot|u_1, u_2, u_3)$)

For the last equation for u_4, we used the fact that $u_3 = C_{3|1,2}^{-1}(w_3|\boldsymbol{u}_{1:2})$ and $u_2 = C_{2|1}^{-1}(w_2|u_1)$ holds by definition.

We turn now to the general case of simulating from C-vine copulas. In this case, the corresponding $d \times d$-dimensional R-vine structure matrix has the following form up to relabeling of the nodes.

$$M = \begin{pmatrix} 1 & 1 & 1 & 1 & 1 & \dots \\ & 2 & 2 & 2 & 2 & \dots \\ & & 3 & 3 & 3 & \dots \\ & & & 4 & 4 & \dots \\ & & & & 5 & \dots \\ & & & & & \dots \end{pmatrix}$$

In the special case of C-vines, using Theorem 4.10, we can express the univariate conditional distribution function $C_{i|1:i-k}(u_i|\boldsymbol{u}_{1:i-k})$ as

$$C_{i|1:i-k}(u_i|\boldsymbol{u}_{1:i-k}) \tag{6.5}$$
$$= \frac{\partial C_{i,i-k;1:i-k-1}\left(C_{i|1:i-k-1}(u_i|\boldsymbol{u}_{1:i-k-1}), C_{i-k|1:i-k-1}(u_{i-k}|\boldsymbol{u}_{1:i-k-1})\right)}{\partial C_{i-k|1:i-k-1}(u_{i-k}|\boldsymbol{u}_{1:i-k-1})}$$

for $i = 1, \dots, d$ and $k = 1, \dots, i-2$. The sampling algorithm stores the copula parameters also in a strict upper triangular $d \times d$ matrix Θ with entries $\eta_{ik} = \theta_{ik;1:i-1}$ for $i < k = 2, \dots, d$ given by

$$\Theta := \begin{pmatrix} - & \theta_{1,2} & \theta_{1,3} & \theta_{1,4} & \dots \\ - & - & \theta_{2,3;1} & \theta_{2,4;1} & \dots \\ - & - & - & \theta_{3,4;1,2} & \dots \\ - & - & - & - & \dots \end{pmatrix}.$$

This matrix Θ is now utilized to calculate the entries of the following upper triangular $d \times d$ matrix V of conditional distribution functions with entries v_{ik} for $i \leq k = 1, \ldots, d$:

$$
V := \begin{pmatrix}
u_1 & u_2 & u_3 & u_4 & \cdots \\
 & C_{2|1}(u_2|u_1) & C_{3|1}(u_3|u_1) & C_{4|1}(u_4|u_1) & \cdots \\
 & & C_{3|1,2}(u_3|u_1, u_2) & C_{4|1,2}(u_4|u_1, u_2) & \cdots \\
 & & & C_{4|1,2,3}(u_4|u_1, u_2, u_3) & \cdots \\
 & & & & \cdots
\end{pmatrix}.
$$

$$(6.6)$$

To determine the entries of V, we first use that $C_{j|1:j-1}(u_j|\boldsymbol{u}_{1:j-1}) = w_j$, i.e., $v_{jj} = w_j$ for $j = 1, \ldots, d$ holds. Here, the variables w_j are defined as in Theorem 6.2. Finally, we rewrite (6.5) in terms of h-functions as

$$
C_{i|1:i-k}(u_i|\boldsymbol{u}_{1:i-k}) = h_{i|i-k;1:i-k-1}\left(C_{i|1:i-k-1}(u_i|\boldsymbol{u}_{1:i-k-1})|C_{i-k|1:i-k-1}(u_{i-k}|\boldsymbol{u}_{1:i-k-1})\right),
$$

and invert with respect to the first argument to obtain for $i = 1, \ldots, d$ and $k = 1, \ldots, i-2$

$$
C_{i|1:i-k-1}(u_i|\boldsymbol{u}_{1:i-k}) \tag{6.7}
$$
$$
= h^{-1}_{i|i-k;1:i-k-1}\left(C_{i|1:i-k}(u_i|\boldsymbol{u}_{1:i-k-1})\big|C_{i-k|1:i-k-1}(u_{i-k}|\boldsymbol{u}_{1:i-k-1})\right).
$$

More precisely, Algorithm 6.4 (taken from Stöber and Czado 2017) applies the relationship (6.7) recursively to calculate the entries v_{ij} of matrix V defined in (6.6). It generates a sample $(u_1, \ldots, u_d)$ from the C-vine copula, which is stored in the first row of the matrix V.

Algorithm 6.4 (*Sampling from a C-vine copula*) The input is a strict upper triangular matrix Θ of copula parameters with entries $\eta_{ki} = \theta_{ki;1:k-1}$ for $k < i$ and $i, k = 1, \ldots, d$ for the d-dimensional C-vine, the output will be a sample from the C-vine copula.

$$\text{Sample } w_i \stackrel{\text{i.i.d.}}{\sim} U[0; 1], \quad i = 1, \ldots, d \tag{1}$$

$$v_{1,1} := w_1 \tag{2}$$

$$\text{FOR } i = 2, \ldots, d \tag{3}$$

$$v_{i,i} := w_i \tag{4}$$

$$\text{FOR } k = i - 1, \ldots, 1 \tag{5}$$

$$v_{k,i} := h^{-1}_{i|k;1:k-1}(v_{k+1,i}|v_{k,k}, \eta_{k,i}) \tag{6}$$

END FOR

END FOR

$$\text{RETURN } u_i := v_{1,i} \quad i = 1, \ldots, d \tag{7}$$

Table 6.1 R-vine specification: Chosen pair copula families, their family name abbreviations in VineCopula, parameter value, and corresponding Kendall's τ value

Edge	Family	Abbrev.	θ	τ
12	Clayton (C)	3	4.80	0.71
13	Gaussian (N)	1	0.50	0.33
14	Gaussian (N)	1	0.90	0.71
15	Gumbel (G)	4	3.90	0.74
23;1	Gumbel (G)	4	1.90	0.47
24;1	Rotated Gumbel 90° (G90)	24	−2.60	−0.62
25;1	Rotated Gumbel 90° (G90)	24	−6.50	−0.85
34;12	Rotated Clayton 90° (C90)	3	−5.10	−0.72
35;12	Clayton (C)	3	0.90	0.31
45;123	Gaussian (N)	1	0.20	0.13

Remark 6.5 It is not needed to store the whole matrix V in this algorithm, since the only entries we are going to use more than once are $v_{i,i} = w_i$. Thus, we can always delete or overwrite the other entries after they have been used as input for the next recursion. The matrix structure is chosen to illustrate the iterative procedure. It is also needed to understand the general regular vine case discussed later.

Ex 6.4 (Simulating from a five-dimensional C-vine copula) The function RVineSim from the R library VineCopula can be used to simulate from C-vine copulas. As inputs we need an RVineMatrix object, which stores structure matrix M, the pair copula families and their parameters. To create such an RVineMatrix object for C-vines, the function C2Rvine can be used. For our example, we choose the pair copula families together with their parameter and the corresponding Kendall's τ specified in Table 6.1. If we fix a Kendall's τ value, we can use the function BiCopTau2Par to determine the corresponding parameter value. This is unique, since we used only pair copula families with a single parameter.

The corresponding C-vine tree sequence together with copula family and Kendall's τ values are given in Fig. 6.2. This is produced using the function plot for an RVineMatrix object within the VineCopula package.

To get an idea, how the pair copula specification of Table 6.1 looks at the normalized contour level, we present these contour plots shown in Fig. 6.3. This is produced with the function contour applied to a RVineMatrix object within the VineCopula package.

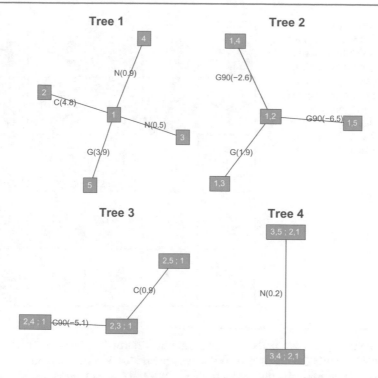

Fig. 6.2 C-vine tree plots: Tree sequence with families and Kendall's τ values

Fig. 6.3 Normalized contours: Normalized theoretical contour plots of all pair copulas specified in the C-vine tree sequence given in Fig. 6.2

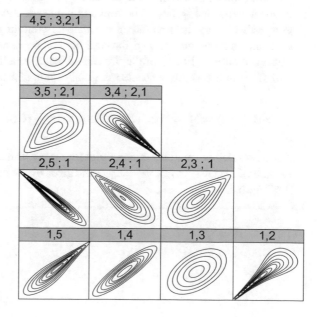

Fig. 6.4 `C-vine`
`simulation`: A simulated
sample of size 1000 from the
C-vine specified in Table 6.1
(upper triangle: pair scatter
plot of copula data, diagonal:
marginal histograms of
copula data, lower triangle:
empirical normalized
contour plots)

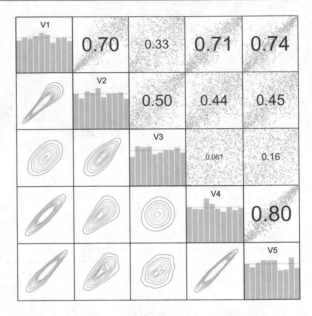

Finally, we present pairs plots and empirical normalized contour plots of a sample from this C-vine of size 1000 shown in Fig. 6.4. We see the typical shapes of the contour plots for the pairs $(1, 2)$, $(1, 3)$, $(1, 4)$, and $(1, 5)$ associated with the pair copulas of the first tree of the C-vine specification (first column of the lower triangle of Fig. 6.4). The remaining empirical normalized contour plots give estimates of the theoretical normalized contour plots. These theoretical plots can only be determined through integration over the appropriate conditioning variables. For example the theoretical normalized contour plot of the pair $(2, 3)$ requires the bivariate marginal copula density

$$c_{23}(u_2, u_3) = \int_0^1 c_{12}(u_1, u_2) c_{13}(u_1, u_3) c_{23;1}(C(u_2|u_1), C(u_3|u_1)) du_1.$$

The associated normalized contour plot for c_{23} is estimated using the normalized sample values $(\Phi^{-1}(u_{i2}), \Phi^{-1}(u_{i3}))$ and given by the panel in row 3 and column 2 of Fig. 6.4.

6.4 Simulating from D-vine Copulas

Now, we consider d-dimensional D-vine copulas with regular vine matrix given as

$$
\begin{pmatrix}
1 & 1 & 2 & 3 & 4 & \dots \\
  & 2 & 1 & 2 & 3 & \dots \\
  &   & 3 & 1 & 2 & \dots \\
  &   &   & 4 & 1 & \dots \\
  &   &   &   & 5 & \dots \\
  &   &   &   &   & \dots
\end{pmatrix}
$$

For sampling from a D-vine copula (right panel of Fig. 5.4), the relationship which we use instead of (6.5) for the C-vine case is

$$
C_{i|k:i-1}(u_i|\boldsymbol{u}_{k:i-1}) = \frac{\partial C_{i,k;k+1:i-1}\left(C_{i|k+1:i-1}(u_i|\boldsymbol{u}_{k+1:i-1}), C_{k|k+1:i-1}(u_k|\boldsymbol{u}_{k+1:i-1})\right)}{\partial C_{k|k+1:i-1}(u_k|\boldsymbol{u}_{k+1:i-1})}, \quad (6.8)
$$

where $i = 3, \dots, d$ and $k = 2, \dots, i-1$, i.e., $i > k$. In contrast to the C-vine copula, we do not automatically obtain the second argument in (6.8). This means that an extra step for its computation has to be added and that we have to determine two upper triangular matrices in $d \times d$ dimensions containing needed conditional distribution functions

$$
V := \begin{pmatrix}
u_1 & u_2 & u_3 & u_4 & \dots \\
    & C_{2|1}(u_2|u_1) & C_{3|2}(u_3|u_2) & C_{4|3}(u_4|u_3) & \dots \\
    &   & C_{3|2,1}(u_3|u_2, u_1) & C_{4|3,2}(u_4|u_3, u_2) & \dots \\
    &   &   & C_{4|3,2,1}(u_4|u_3, u_2, u_1) & \dots \\
    &   &   &   & \dots
\end{pmatrix}
$$

$$(6.9)$$

$$
V^2 := \begin{pmatrix}
u_1 & u_2 & u_3 & u_4 & \dots \\
    & C_{1|2}(u_1|u_2) & C_{2|3}(u_2|u_3) & C_{3|4}(u_3|u_4) & \dots \\
    &   & C_{1|2,3}(u_1|u_2, u_3) & C_{2|4,3}(u_2|u_4, u_3) & \dots \\
    &   &   & C_{1|4,3,2}(u_1|u_4, u_3, u_2) & \dots \\
    &   &   &   & \dots
\end{pmatrix}
$$

$$(6.10)$$

using the strict upper triangular matrix Θ with entries η_{ki} for $k < i$

$$
\Theta := \begin{pmatrix}
- & \theta_{1,2} & \theta_{2,3} & \theta_{3,4} & \dots \\
- & - & \theta_{3,1;2} & \theta_{4,2;3} & \dots \\
- & - & - & \theta_{4,1;3,2} & \dots \\
- & - & - & - & \dots
\end{pmatrix}.
$$

$$(6.11)$$

In particular, we have

$$\eta_{ki} = \theta_{i-k,i;i-k+1:i-1}$$

for $k = 1, \ldots, d-1$ and $i = k+1, \ldots, d$. Recall that the order of the two conditioned indices is not important, i.e., $\theta_{e_a,e_b;D_e} = \theta_{e_b,e_a;D_e}$ holds. Before we present the general algorithm, we illustrate it in four dimensions.

Ex 6.5 (*Simulating from a four-dimensional D-vine copula*) We will use the matrices V, V^2, and Θ as defined in (6.9), (6.10) and (6.11) for the special case when $d = 4$. To simulate from the four-dimensional D-vine copula we need to calculate the following conditional distributions:

$$C_{4|1:3}(u_4|\boldsymbol{u}_{1:3}) = h_{4|1;2:3}(C_{4|2:3}(u_4|\boldsymbol{u}_{2:3})|C_{1|2:3}(u_1|\boldsymbol{u}_{2:3}); \theta_{14;23}) \qquad (6.12)$$

$$C_{3|1:2}(u_3|\boldsymbol{u}_{1,2}) = h_{3|1;2}(C_{3|2}(u_3|u_2)|C_{1|2}(u_1|u_2); \theta_{13;2}) \qquad (6.13)$$

$$C_{2|1}(u_2|u_1) = h_{2|1}(u_2|u_1; \theta_{12}).$$

Now, we sample w_1, w_2, w_3, and w_4 independently and identically from the uniform distribution on the unit interval and set $u_1 := w_1, u_2 := C_{2|1}^{-1}(w_2|u_1), u_3 := C_{3|1:2}^{-1}(w_3|\boldsymbol{u}_{1:2})$ and $u_4 := C_{4|1:3}(w_4|\boldsymbol{u}_{1:3})$. Now (u_1, u_2, u_3, u_4) is a realization from the four-dimensional D-vine copula. Further, we note that $w_1 = u_1$, $w_2 = C_{2|1}(u_2|u_1)$, $w_3 = C_{3|1:2}(u_3|\boldsymbol{u}_{1:2})$ and $w_4 = C_{4|1:3}(u_4|\boldsymbol{u}_{1:3})$ holds.

Inverting the right-hand side of the equations (6.12) and (6.13) with respect to the first argument gives

$$C_{3|2}(u_3|u_2) = h_{3|1;2}^{-1}(w_3|C_{1|2}(u_1|u_2); \theta_{13;2}) \qquad (6.14)$$

$$C_{4|2:3}(u_4|\boldsymbol{u}_{2:3}) = h_{4|1;2:3}^{-1}(w_4|C_{1|2:3}(u_1|\boldsymbol{u}_{2:3}); \theta_{14;23}). \qquad (6.15)$$

We now determine the entries of V and V^2. For the first column of V and V^2, we set

$$v_{11} := w_1 = u_1 \text{ and } v_{11}^2 := w_1 = u_1.$$

For the second column of V and V^2, we define

$$v_{22} := w_2 = C_{2|1}(u_2|u_1)$$

$$v_{12} := u_2 = C_{2|1}^{-1}(w_2|u_1) = h_{2|1}^{-1}(v_{22}|v_{11}^2, \eta_{12})$$

$$v_{12}^2 := u_2$$

$$v_{22}^2 := C_{1|2}(u_1|u_2) = h_{1|2}(u_1|u_2, \theta_{12}) = h_{1|2}(v_{11}^2|v_{12}, \eta_{12})$$

Now, we proceed with the third column of V and V^2 as follows:

$$v_{33} := w_3 = C_{3|1:2}(u_3|\boldsymbol{u}_{1:2})$$
$$v_{23} := C_{3|2}(u_3|u_2) = h^{-1}_{3|1:2}(w_3|C_{1|2}(u_1|u_2), \theta_{13;2}) = h^{-1}_{3|1:2}(v_{33}|v_{22}^2, \eta_{23})$$
$$v_{33}^2 := C_{1|2:3}(u_1|\boldsymbol{u}_{2:3}) = h_{1|3;2}(C_{1|2}(u_1|u_2)|C_{3|2}(u_3|u_2); \theta_{13;2}) = h_{1|3;2}(v_{22}^2|v_{23}; \eta_{23})$$
$$v_{13} := u_3 = C^{-1}_{3|1:2}(w_3|\boldsymbol{u}_{1:2}) = h^{-1}_{3|2}(h^{-1}_{3|1:2}(w_3|C_{1|2}(u_1|u_2); \theta_{13;2})|u_2; \theta_{23}) = h^{-1}_{3|2}(v_{23}|v_{12}^2; \eta_{13})$$
$$v_{13}^2 := v_{13}$$
$$v_{23}^2 := C_{2|3}(u_2|u_3) = h_{2|3}(v_{12}^2|v_{13}; \eta_{13}).$$

For the entry v_{23}, we used (6.14) and for the entry v_{13} equation (6.13) with inversion. For the entry v_{33}^2, we used Eq. (4.21).

For the fourth column of V, we get:

$$v_{44} := w_4 = C_{4|1:3}(u_4|\boldsymbol{u}_{1:3})$$
$$v_{34} := C_{4|2:3}(u_4|\boldsymbol{u}_{2:3}) = h^{-1}_{4|1;2:3}(w_4|C_{1|2:3}(u_1|\boldsymbol{u}_{2:3}); \theta_{14;23}) = h^{-1}_{4|1;2:3}(v_{44}|v_{33}^2, \eta_{34})$$
$$v_{24} := C_{4|3}(u_4|u_3) = h^{-1}_{4|2;3}(C_{4|23}(u_4|\boldsymbol{u}_{2:3})|C_{2|3}(u_2|u_3); \theta_{24;3}) = h^{-1}_{4|2;3}(v_{34}|v_{23}^2; \eta_{24})$$
$$v_{14} := u_4 = C^{-1}_{4|1:3}(w_4|\boldsymbol{u}_{1:3}) = h^{-1}_{4|3}(C_{4|3}(u_4|u_3)|u_3; \theta_{34}) = h^{-1}_{4|3}(v_{24}|v_{13}^2; \eta_{14}).$$

(6.16)

For v_{34}, we used (6.15) and for v_{24}, we used a similar calculation as for v_{23}. Finally, v_{14} requires the inversion of (6.12) and the use of previously defined elements of V and V^2. For sampling from this four-dimensional D-vine copula, only the values v_{ii} for $i = 1, \ldots, 4$ are required, the remaining values of the fourth column of V are just given for completeness.

As we see, we need to evaluate the second argument of $C_{4|1:3}$ separately. This is also intuitively clear since for a D-vine the conditioning sets in each tree T_j is changing for each edge in the tree, while this is not the case for C-vines.

The calculations of the entries of V and V^2 in arbitrary dimension can be facilitated using the recursions given in Algorithm 6.6. It is a slightly adapted version of the algorithm contained in Stöber and Czado (2017).

Algorithm 6.6 (*Sampling from a D-vine copula*) The input of the algorithm is the matrix Θ specified in (6.11) of copula parameters for the d-dimensional D-vine, the output will be a sample from the D-vine distribution.

Sample $w_i \overset{i.i.d.}{\sim} U[0; 1], \quad i = 1, \ldots, d$ (1)

$v_{1,1} := w_1, v_{1,1}^2 := w_1$ (2)

FOR $i = 2, \ldots, d$ (3)

$\quad v_{i,i} := w_i$ (4)

$\quad$ FOR $k = i - 1, \ldots, 1$ (5)

$\quad\quad v_{k,i} := h^{-1}_{i|i-k;i-k+1:i-1}(v_{k+1,i}|v_{k,i-1}^2, \eta_{k,i})$ (6)

$$\text{IF } i < d \tag{7}$$

$$v^2_{k+1,i} := h_{i-k|i;i-k+1:i-1}(v^2_{k,i-1}|v_{k,i}, \eta_{k,i}) \tag{8}$$

END IF

END FOR

$$v^2_{1,i} := v_{1,i} \tag{9}$$

END FOR

$$\text{RETURN } v_i := v_{1,i} \quad i = 1, \ldots, d \tag{10}$$

(Taken from Stöber and Czado 2017. With permission of ©World Scientific Publishing Co. Pte. Ltd. 2017.)

Remark 6.7 Just as we noted for the C-vine copula in Remark 6.5, we do not need to store all matrix entries calculated during the recursion. The entries in row i of the first matrix V can always be deleted after they have been used to calculate u_i and the entries of the second matrix V^2.

6.5 Simulating from Regular Vine Copulas

After the construction of simulation algorithms for C- and D-vine copulas, we develop now a *simulation algorithm from a general R-vine copula*. The C- and D-vine tree sequences are extreme cases. In the C-vine case, there is one root node in each tree, which shares edges with all other nodes, while in the D-vine case each node has edges at most with *two* other nodes. In the simulation algorithms, we need a set of conditional distribution functions corresponding to the first argument (stored in matrix V) and the second argument (stored in the matrix V^2) for the inverse h-function. For the C-vine case, the entries for the second argument are already determined by the entries of the matrix V. For the general R-vine copula, the sampling procedure will be a combination between these two extreme cases and will depend on the number of edges for each node.

Choose now a d-dimensional R-vine matrix $M = (m_{i,j})_{i,j=1,\ldots,d}$ as defined in Sect. 5.5. Without loss of generality, we assume that the entries on the diagonal are ordered as $1, 2, \ldots, d$, i.e., $m_{i,i} = i$. If this is not the case use a permutation of the indices $1, 2, \ldots, d$. The parameters for the pair copulas associated with M are stored in a strict upper triangular $d \times d$ matrix Θ. Here, the (k, i)th entry $\eta_{k,i}$ for $k < i$ of Θ corresponds to the parameter of $c_{m_{k,i}, m_{k,k}; m_{1,i}, \ldots, m_{k-1,i}}$. More precisely define the matrix Θ as

$$\Theta = \begin{pmatrix} - & \theta_{m_{1,2},2} & \theta_{m_{1,3},3} & \theta_{m_{1,4},4} & \cdots \\ - & - & \theta_{m_{2,3},3;m_{1,3}} & \theta_{m_{2,4},4;m_{1,4}} & \cdots \\ - & - & - & \theta_{m_{3,4},4;m_{2,4},m_{1,4}} & \cdots \\ - & - & - & - & \cdots \end{pmatrix}. \tag{6.17}$$

In contrast to C- and D-vine copulas different shapes of the associated R-vine trees T_j are possible for R-vine copulas. In this more general case, we still need to express all required arguments for the h function in terms of the R-vine matrix M. More precisely, the conditional distribution function corresponding to $U_{m_{i,i}}$ can be determined by $U_{m_{k,i}} = u_{m_{k,i}}, \ldots, U_{m_{1,i}} = u_{m_{1,i}}$ (compare to (6.5) and (6.8)) as

$$
C(u_{m_{i,i}}|u_{m_{k,i}}, u_{m_{k-1,i}} \cdots, u_{m_{1,i}}) =
$$
$$
\frac{\partial C_{m_{i,i},m_{k,i};m_{k-1,i},\ldots,m_{1,i}} \left(C(u_{m_{i,i}}|u_{m_{k-1,i}}, \ldots, u_{m_{1,i}}), C(u_{m_{k,i}}|u_{m_{k-1,i}}, \ldots, u_{m_{1,i}}) \right)}{\partial C(u_{m_{k,i}}|u_{m_{k-1,i}}, \ldots, u_{m_{1,i}})}
$$
(6.18)

for $i = 3, \ldots, d$ and $k = 2, \ldots, i - 1$. To shorten notation, we left out the sub indices of the conditional distribution functions. Again, we define two upper triangular matrices V and V^2 as follows:

$$
V = \begin{pmatrix} u_1 & u_2 & u_3 & u_4 & \ldots \\ & C(u_2|u_{m_{1,2}}) & C(u_3|u_{m_{1,3}}) & C(u_4|u_{m_{1,4}}) & \ldots \\ & & C(u_3|u_{m_{1,3}}, u_{m_{2,3}}) & C(u_4|u_{m_{1,4}}, u_{m_{2,4}}) & \ldots \\ & & & C(u_4|u_{m_{1,4}}, u_{m_{2,4}}, u_{m_{3,4}}) & \ldots \\ & & & & \ldots \end{pmatrix}
$$
(6.19)

$$
V^2 = \begin{pmatrix} u_1 & u_2 & u_3 & u_4 & \ldots \\ & C(u_{m_{1,2}}|u_2) & C(u_{m_{1,3}}|u_3) & C(u_{m_{1,4}}|u_4) & \ldots \\ & & C(u_{m_{2,3}}|u_{m_{1,3}}, u_3) & C(u_{m_{2,4}}|u_{m_{1,4}}, u_4) & \ldots \\ & & & C(u_{m_{3,4}}|u_{m_{1,4}}, u_{m_{2,4}}, u_4) & \ldots \\ & & & & \ldots \end{pmatrix}.
$$
(6.20)

The matrix V^2 is similarly defined as the corresponding one for the D-vine copula.

Two questions have to be addressed before a sampling algorithm can be constructed: From which column of the R-vine matrix do we have to select the second argument of the inverse h-function (and the first argument of the h-function)? Do we have to select it from the first matrix V defined in (6.19) or from the second matrix V^2 defined in (6.20)?

For the answer to these questions, we consider the R-vine matrices M corresponding to the C-vine and D-vine copula given in (5.6) and (5.7), respectively. Using these we learn how the choices in Algorithms 6.4 and 6.6 can be expressed in terms of R-vine matrices. This will give us an idea of how to choose the right arguments for a general R-vine copula. Additionally, we introduce the matrix $\mathcal{M} = (\tilde{m}_{k,i}), k \leq i$, with elements

$$
\tilde{m}_{k,i} := max\{m_{k,i}, m_{k-1,i}, \ldots, m_{1,i}\}.
$$

For the C-vine copula with matrix M we have $\mathcal{M} = M$, i.e., we have $\tilde{m}_{k,i} = k$ for all entries of $\mathcal{M}$. However for the D-vine copula matrix M, we have $\mathcal{M} \neq M$. In particular $\tilde{m}_{k,i} = i - 1$ holds for all off-diagonal entries of $\mathcal{M}$.

For the C-vine copula, we always select an entry for the second argument for the inverse h-function from the kth column of V within Step (6) of Algorithm 6.4.

This corresponds to column $m_{k,i} = \tilde{m}_{k,i} = k$. For the D-vine copula, we remain in column $i - 1$ ($= \tilde{m}_{k,i}$) in Step (6) of Algorithm 6.6 for the second argument of the inverse h-function. Therefore, the sampling algorithm for an R-vine copula will always choose $\tilde{m}_{k,i}$. The entry, which is needed as second argument for the inverse h-function *has* to be in this column, since the second argument in (6.18) is $C(u_{m_{k,i}} | u_{m_{k-1,i}}, \ldots, u_{m_{1,i}})$, and $\tilde{m}_{k,i}$ is the largest index in this expression. The first row where an entry containing index $\tilde{m}_{k,i}$ can be located is row $\tilde{m}_{k,i}$, since the diagonals of M are arranged in increasing order and since $m_{l,h} \leq h$ for $l = 1, \ldots, h$ by Property 3 of Definition 5.23. Furthermore, each element in column h of (6.19) and (6.20) contains the index h, which means that the entry we are looking for cannot be found in a column to the right of column $\tilde{m}_{k,i}$.

In matrix V of (6.19), all entries in column $\tilde{m}_{k,i}$ are conditional distribution functions of $U_{\tilde{m}_{k,i}}$ given other variables, and in matrix V^2 defined in (6.20) $U_{\tilde{m}_{k,i}}$ is part of the conditioned variables. Thus, we only need to check whether $\tilde{m}_{k,i} = m_{k,i}$ to choose from the appropriate matrix.

Algorithm 6.8 taken in slightly corrected form from Stöber and Czado (2017) summarizes the results of the discussion. Induction can be used to proof that at each step of the algorithm the appropriate entry can selected from two matrices V and V^2 with the form given in (6.19) and (6.20), respectively. A more formal proof can be found in Chap. 5 of Dißmann (2010).

Algorithm 6.8 (*Sampling from an R-vine copula*) The input of the algorithm is a matrix Θ with entries $\eta_{k,i}$, given in (6.17), of copula parameters for the d-dimensional R-vine copula. The output will be a sample from the R-vine copula.

$$\text{Sample } w_i \overset{\text{i.i.d.}}{\sim} U[0; 1], \quad i = 1, \ldots, d \tag{1}$$

$$v_{1,1} := w_1 \tag{2}$$

$$\text{FOR } i = 2, \ldots, d \tag{3}$$

$$\quad v_{i,i} := w_i \tag{4}$$

$$\quad \text{FOR } k = i - 1, \ldots, 1 \tag{5}$$

$$\quad\quad \text{IF } (m_{k,i} = \tilde{m}_{k,i}) \tag{6}$$

$$\quad\quad\quad v_{k,i} := h^{-1}_{m_{ii} | m_{ki}; m_{1i}, \ldots m_{k-1,i}} (v_{k+1,i} | v_{k,\tilde{m}_{k,i}}, \eta_{k,i}) \tag{7a}$$

$$\quad\quad \text{ELSE}$$

$$\quad\quad\quad v_{k,i} := h^{-1}_{m_{ii} | m_{ki}; m_{1i}, \ldots m_{k-1,i}} (v_{k+1,i} | v^2_{k,\tilde{m}_{k,i}}, \eta_{k,i}) \tag{7b}$$

$$\quad\quad \text{END IF ELSE}$$

$$\quad\quad \text{IF } (i < d) \tag{8}$$

$$\quad\quad\quad \text{IF } (m_{k,i} = \tilde{m}_{k,i}) \tag{9}$$

$$\quad\quad\quad\quad v^2_{k+1,i} := h_{m_{ki} | m_{ii}; m_{1i}, \ldots m_{k-1,i}} (v_{k,\tilde{m}_{k,i}} | v_{k,i}, \eta_{k,i}) \tag{10a}$$

$$\quad\quad\quad \text{ELSE}$$

$$v^2_{k+1,i} := h_{m_{ki}|m_{ii};m_{1i},\dots m_{k-1,i}}(v^2_{k,\tilde{m}_{k,i}}|v_{k,i}, \eta_{k,i}) \quad (10b)$$

      END IF ELSE

    END IF

  END FOR

END FOR

RETURN $u_i := v_{1,i}$ $i = 1,\dots,d$ (11)

(Taken from Stöber and Czado 2017. With permission of ©World Scientific Publishing Co. Pte. Ltd. 2017.)

Ex 6.6 (Sampling from the regular vine of Example 5.6 restricted to five dimensions) To illustrate the steps of Algorithm 6.8, we consider Example 5.6 restricted to five dimensions to shorten our exposition. For this, we remove the nodes 6, 56, 16; 5, 36; 15, and 46; 135 from the trees $T_1, \dots, T_5$, respectively. The corresponding R-vine matrix M is given by (5.5) with the sixth row and column removed. We now reorder the diagonal in increasing order giving the following matrices:

$$M = \begin{pmatrix} 1 & 1 & 2 & 3 & 3 \\ & 2 & 1 & 2 & 2 \\ & & 3 & 1 & 1 \\ & & & 4 & 4 \\ & & & & 5 \end{pmatrix} \text{ and } \tilde{M} = \begin{pmatrix} 1 & 1 & 2 & 3 & 3 \\ & 2 & 2 & 3 & 3 \\ & & 3 & 3 & 3 \\ & & & 4 & 4 \\ & & & & 5 \end{pmatrix},$$

respectively. Further, we have the matrix V and V^2 given as

$$V = \begin{pmatrix} u_1 & u_2 & u_3 & u_4 & u_5 \\ & C(u_2|u_1) & C(u_3|u_2) & C(u_4|u_3) & C(u_5|u_3) \\ & & C(u_3|u_2,u_1) & C(u_4|u_3,u_2) & C(u_5|u_3,u_2) \\ & & & C(u_4|u_3,u_2,u_1) & C(u_5|u_3,u_2,u_1) \\ & & & & C(u_5|u_3,u_2,u_1,u_4) \end{pmatrix} \quad (6.21)$$

$$V^2 = \begin{pmatrix} u_1 & u_2 & u_3 & u_4 & u_5 \\ & C(u_1|u_2) & C(u_2|u_3) & C(u_3|u_4) & C(u_3|u_5) \\ & & C(u_1|u_2,u_3) & C(u_2|u_3,u_4) & C(u_2|u_3,u_5) \\ & & & C(u_1|u_3,u_2,u_4) & C(u_1|u_3,u_2,u_5) \\ & & & & C(u_4|u_3,u_2,u_1,u_5) \end{pmatrix}. \quad (6.22)$$

Finally, the matrix Θ of pair copula parameter is given by

$$\Theta = \begin{pmatrix} - & \theta_{12} & \theta_{23} & \theta_{34} & \theta_{35} \\ - & - & \theta_{13;2} & \theta_{24;3} & \theta_{25;3} \\ - & - & - & \theta_{14;23} & \theta_{15;23} \\ - & - & - & - & \theta_{45;123} \end{pmatrix}. \quad (6.23)$$

For columns 1–4 of M corresponding to $u_1, \ldots, u_4$, this matrix is the same as for a D-vine copula, which means that except for row 1, we have $m_{k,i} \neq \tilde{m}_{k,i}$ and that we select the second entry of the inverse h-function from the second matrix V^2 (6.20). In row 1, $m_{1,i} = \tilde{m}_{1,i}$ for $i = 1, \ldots, 5$ such that in the last step of the iteration for $u_1, \ldots, u_4$ we select from (6.19).

To obtain $u_5 (= v_{15})$, we calculate $C(u_5 | u_1, u_2, u_3) (= v_{45})$ from $C(u_5 | u_1, u_2, u_3, u_4) = w_5 (= v_{55})$ as

$$C(u_5 | u_1, u_2, u_3)(= V_{45}) = h_{5|4;123}^{-1} \left(w_5 (= v_{55}) | C(u_4 | u_1, u_2, u_3)(= v_{44}), \theta_{45|123}(= \eta_{45}) \right).$$

Since $C(u_4 | u_1, u_2, u_3) = v_{44}$ is given in Matrix (6.21) and $m_{45} = 4 = \tilde{m}_{45}$ it gets correctly selected. For the next two steps of the recursion, we need $C(u_1 | u_2, u_3)(= v_{33}^2)$ and $C(u_2 | u_3)(= v_{32}^2)$, which are given in the third column of the matrix V^2 (6.22). Correspondingly, we have $m_{35} = 1 \neq \tilde{m}_{35} = 3$ and $m_{25} = 2 \neq \tilde{m}_{25} = 3$. In the last step, we do select $u_3 (= v_{13})$ from the third column of matrix V (6.21), $\tilde{m}_{1,5} = 3$.

Remark 6.9 As it was noted in Remark 5.27, an R-vine matrix corresponds to a directed vine, when the diagonal element is arranged as the first argument. This requirement might result in an ordering of the conditioned and conditional indices which are not increasing. However, following this specification, we can introduce one further dimension in the parameter of the h-function giving the angle of rotation, and still apply Algorithm 6.8.

The simplified algorithms for the C-vine and D-vine copulas correspond to the special structure in the corresponding R-vine matrices shown in Example 5.13.

Remark 6.10 (Sampling from R-vine distributions) So far, we only have discussed the simulation from an R-vine copula. For a general R-vine distribution with marginal distribution and quantile functions F_j and F_j^{-1}, we can easily transform a sample $(u_1, \ldots, u_d)$ from the associated R-vine copula to a sample from the R-vine distribution by setting $x_j := F_j^{-1}(u_j)$ for $j = 1, \cdots, d$.

Ex 6.7 (Sampling from a specified R-vine copula) We can use the function `RVineSim` from the R library `VineCopula` to simulate from a specified R-vine copula. As specification for the R-vine tree sequence, we use the following R-vine matrix in lower triangular form as discussed in Example 5.14:

$$
\begin{pmatrix}
2 & 0 & 0 & 0 & 0 \\
5 & 3 & 0 & 0 & 0 \\
3 & 5 & 4 & 0 & 0 \\
1 & 1 & 5 & 5 & 0 \\
4 & 4 & 1 & 1 & 1
\end{pmatrix}.
$$

As pair copula families, we chose

$$
\begin{pmatrix}
0 & 0 & 0 & 0 & 0 \\
1 & 0 & 0 & 0 & 0 \\
3 & 23 & 0 & 0 & 0 \\
24 & 24 & 4 & 0 & 0 \\
4 & 1 & 1 & 3 & 0
\end{pmatrix},
$$

where 1 corresponds to a Gaussian (N), 3 to a Clayton (C), 4 to a Gumbel (G), 23 to rotated 90° Clayton (C90), and 24 to a rotated 90° Gumbel (G90). The corresponding copula parameters were set to

$$
\begin{pmatrix}
0 & 0 & 0 & 0 & 0 \\
0.2 & 0 & 0 & 0 & 0 \\
0.9 & -5.1 & 0 & 0 & 0 \\
-6.5 & -2.6 & 1.9 & 0 & 0 \\
3.9 & 0.9 & 0.5 & 4.8 & 0
\end{pmatrix}
$$

The associated R-vine tree sequence plot is given in Fig. 6.5 and the associated theoretical normalized contour plots of the specified (conditional) pair copulas are given in Fig. 6.6.

Pairs plots and normalized contour plots of a sample of size 1000 using the function RVineSim of the library VineCopula are given in Fig. 6.7. Note that, not all bivariate distributions of pairs are directly specified through the R-vine. For example a theoretical normalized contour plot for the variable pair (2, 5) would require a three-dimensional integration over the variables 1, 3 and 4, it however can be estimated using the 1000 sampled values for the variables 2 and 5. The function RVineSim requires an RVM object, which can be generated by the function RVineMatrix.

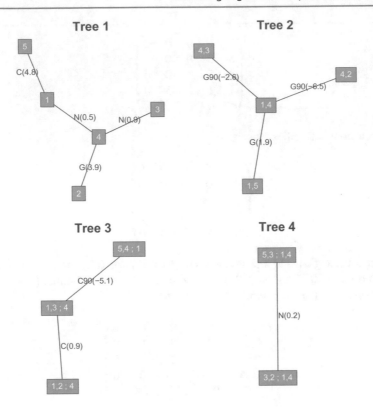

Fig. 6.5 R-vine tree plots: with copula families and Kendall's τ values

Fig. 6.6 Normalized
contours: Theoretical
normalized contour plots of
the specified pair copulas

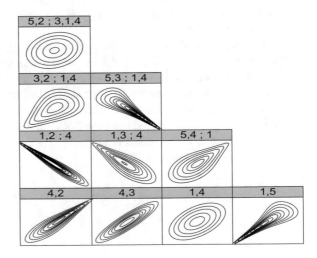

Fig. 6.7 R-vine simulation: Pairwise scatter plots (upper triangular), marginal histograms (diagonal), and pairwise normalized contour plots (lower triangular) of 1000 simulated realizations of the R-vine copula specified in Example 6.7

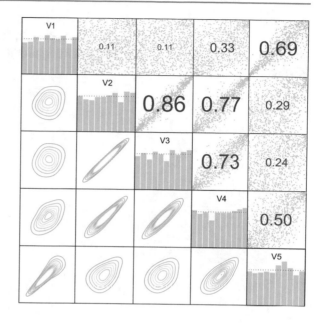

6.6 Exercises

Exer 6.1

(*Recursion for simulation from D-vine copulas*) Proof in detail Equality (6.16) in Example 6.5.

Exer 6.2

(*Simulate from a D vine*) Simulate from a D-vine copula, where the families and parameters are specified as in Table 6.1 of Example 6.4 for the edges of a corresponding five- dimensional D-vine. More precisely, the edges 12, 13, 14 and 15 of the C-vine correspond to the edges 12, 23, 34 and 45 of the D-vine vine. Make a similar correspondence for trees T_2 until tree T_4 to completely specify the D-vine copula. Produce a tree sequence plot and pairs/normalized contour plots for sample of size 1000 from this D-vine copula. Compare the results to ones obtained in Example 6.4.

Exer 6.3

(*Order of the C-vine copula*) Consider you have the variables "A", "B", "C", "D", and "E" available. Simulate from a C-vine with specification as in Example 6.4. Change then only the order of the C-vine to "B", "D", "E", "A", and "C", compare then the two samples of the C-vine with appropriate C-vine tree sequence plots and pairs/contour plots. The plots should have the variables names "A", "B", "C", "D", and "E".

Exer 6.4

(*Verification of general R-vine sampling Algorithm* 6.8) For Example 6.6, determine all entries of the Algorithm 6.8 and verify that they produce a sample of the specified R-vine.

Parameter Estimation in Simplified Regular Vine Copulas

7

In the last chapter, we have seen how to design very flexible multivariate copulas. To make them useful for practise, we have now to tackle the problem of parameter estimation. For this, we will assume that the simplifying assumption holds throughout. We also suppose, that the vine copula model is completely specified by a given vine tree sequence, as well as, the pair copula families associated with each edge in the vine tree sequence. The problem of how to choose the pair copula families and how to select the vine structure will be treated in subsequent chapters.

We study now the estimation of the associated pair copula parameters for pair copulas in a given vine structure. This estimation problem can be decomposed in a sequence of smaller estimation problems, which can be solved efficiently. This sequential solution can be utilized as a starting value for determining joint maximum likelihood.

7.1 Likelihood of Simplified Regular Vine Copulas

For the following, we assume that we have an i.i.d. d-dimensional sample of size n from a regular vine copula with specified vine tree sequence and pair copula families available. We collect this sample in a $n \times d$ data matrix defined as

$$\boldsymbol{u} := (\boldsymbol{u}_1^\top, ..., \boldsymbol{u}_n^\top) \text{ with } \boldsymbol{u}_k := (u_{k,1}, ..., u_{k,d})^\top \text{ for } k = 1, ..., n.$$

Recall from (5.1) that we abbreviate an edge $e = e_{C_{e,a}, C_{e,b}; D_e}$ from the edge set $E = (E_1, \ldots, E_{d-1})$ by

$$e = (a_e, b_e; D_e) \text{ with } e \in E_j \text{ for } j = 1, ..., d - 1.$$

© Springer Nature Switzerland AG 2019
C. Czado, *Analyzing Dependent Data with Vine Copulas*, Lecture Notes
in Statistics 222, https://doi.org/10.1007/978-3-030-13785-4_7

Using the expression for the regular vine density given in (5.2), we can easily determine the likelihood of a regular vine copula under the simplifying assumption.

Definition 7.1 (*Likelihood of a simplified regular vine copula*) The likelihood of a simplified regular vine copula with copula parameters $\theta = \{\theta_e, e \in E\}$ and observed data u is defined as

$$\ell(\theta; u) = \prod_{k=1}^{n} \prod_{j=1}^{d-1} \prod_{e \in E_j} c_{a_e,b_e;D_e}(C_{a_e|D_e}(u_{k,a_e}|u_{k,D_e}), C_{b_e|D_e}(u_{k,b_e}|u_{k,D_e})).$$

$$(7.1)$$

Remark 7.2 In the likelihood expression (7.1), we suppressed the dependence on the associated copula parameters. In particular, the pair copula $c_{a_e,b_e;D_e}$ depends on $\theta_{a_e,b_e;D_e}$, while the arguments $C_{a_e|D_e}$ and $C_{b_e|D_e}$ depend on the parameters of the pair copulas, which are needed to recursively determine them. It is important to note, that the recursion only uses pair copulas, which are identified by the vine tree sequence. In Example 5.10, we see from Fig. 5.5 that $C_{5|1,3,4}$ is determined by $c_{5,4;1,3}$, $c_{5,3;1}$, $c_{4,1;3}$, $c_{5,1}$, $c_{3,1}$, and $c_{4,3}$ and thus, it depends on the parameters associated with these terms.

We now discuss the special case in three dimensions, as well as the C-vine and the D-vine case in arbitrary dimension.

Ex 7.1 (*Likelihood in a three-dimensional PCC*) Using the copula density (4.10), the likelihood for $\theta = (\theta_{12;3}, \theta_{23}, \theta_{12})^\top$ can be expressed as

$$\ell(\theta; u) = \prod_{k=1}^{n} c_{13;2}(C_{1|2}(u_{k,1}|u_{k,2}; \theta_{12}), C_{3|2}(u_{k,1}|u_{k,2}; \theta_{23}); \theta_{13;2})$$
$$\times c_{23}(u_{k,2}, u_{k,3}; \theta_{23})c_{12}(u_{k,1}, u_{k,2}; \theta_{12}). \qquad (7.2)$$

Ex 7.2 (*Likelihood of a parametric D-vine and C-vine copula*) We assume that all pair copula families needed for a C-vine or D-vine copula are parametric. In particular for a D-vine, the parameter(s) of the pair copula $C_{i,i+j;i+1:i+j-1}$ are denoted by $\theta_{i,i+j;i+1:i+j-1}$ for $j = 1, \ldots d - 1$ and $i = 1, \ldots, d - j$, and all parameters are collected in θ. Correspondingly, we use $\theta_{j,i+j;1:j-1}$ for the parameter(s) of the pair copula $C_{j,i+j;1:j-1}$ and again collect all parameters in θ.

Using (4.18) and (4.20), respectively, the *likelihood of the D-vine copula* is given by

$$\ell(\boldsymbol{\theta}, \boldsymbol{u}) = \prod_{k=1}^{n} \prod_{j=1}^{d-1} \prod_{i=1}^{d-j} \tag{7.3}$$

$$\times \; c_{i,i+j;i+1:i+j-1}(C_{i|i+1:i+j-1}(u_{k,i}|\boldsymbol{u}_{k,i+1:i+j-1}), C_{i+j|i+1:i+j-1}(u_{k,i+j}|$$

$$\boldsymbol{u}_{k,i+1:i+j-1})),$$

while the *likelihood of a C-vine copula* can be written as

$$\ell(\boldsymbol{\theta}, \boldsymbol{u}) = \prod_{k=1}^{n} \prod_{j=1}^{d-1} \prod_{i=1}^{d-j} \tag{7.4}$$

$$\times \; c_{j,i+j;1:j-1}(C_{i|1:j-1}(u_{k,i}|\boldsymbol{u}_{k,1:j-1}), C_{i+j|1:j-1}(u_{k,i+j}|\boldsymbol{u}_{k,1:j-1})).$$

In particular for D-vine copulas, the conditional distribution function $C_{i|i+1:i+j-1}$ depends first on $\boldsymbol{\theta}_{i,i+j-1;i+j-2}$, while $C_{i+j|i+1:i+j-1}$ on $\boldsymbol{\theta}_{i+2,i+j-1;i+j-1}$. This can be used recursively to determine the complete parameter dependence.

7.2 Sequential and Maximum Likelihood Estimation in Simplified Regular Vine Copulas

The standard parameter estimation method in statistics is maximum likelihood. This involves maximizing the likelihood. Often, it is easier to maximize the log-likelihood instead of the likelihood. Here it is particularly easy, since the likelihood of a regular vine copula is a product of terms, which transforms to a sum, when we optimize the log-likelihood. For hierarchical Archimedean copulas (see for example Savu and Trede 2010 and Okhrin et al. 2013), this is not possible.

In the case of simplified regular vine copulas in d dimensions with pair copulas with a single parameter, the number of parameters to be maximized is $d(d-1)/2$, which grows quadratically. Therefore it is important to find good starting values for the optimization. Such starting values are given by sequential parameter estimates. We first illustrate this approach in the three dimensional case.

Ex 7.3 (Sequential estimation in three dimensions) The associated likelihood for the three-dimensional case is given in (7.2). In the first step, we find parameter estimates of $\boldsymbol{\theta}_{12}$ and $\boldsymbol{\theta}_{23}$ by using the bivariate sub samples $(u_{k,1}, u_{k,2})$ and

$(u_{k,2}, u_{k,3})$ for $k = 1, ..., n$, respectively. Here, we can use maximum likelihood or the inversion of Kendall's τ, when pair copula families with a single parameter have been used (see Sect. 3.10). We denote the corresponding estimates by $\hat{\theta}_{12}$ and $\hat{\theta}_{23}$, respectively.

In the second step, we define the pseudo-observations

$$u_{k,1|2,\hat{\theta}_{12}} := C_{1|2}(u_{k,1}|u_{k,2}, \hat{\theta}_{12})$$
$$u_{k,3|2,\hat{\theta}_{23}} := C_{3|2}(u_{k,3}|u_{k,2}, \hat{\theta}_{23}) \tag{7.5}$$

for $k = 1, ..., n$. Under the simplifying assumption, these provide an approximate i.i.d. sample from the pair copula $C_{13;2}$. Further, the marginal distribution associated with the pseudo-observations (7.5) is approximately uniform, since the transformation in (7.5) is a probability integral transform with estimated parameter values (compared to the definition of CPIT given in (3.32)). Therefore, we use the pseudo-observations to estimate the parameter(s) of the pair copula $c_{13;2}$ by again either maximizing

$$\prod_{k=1}^{n} c_{13;2}(u_{k,1|2,\hat{\theta}_{12}}, u_{k,3|2,\hat{\theta}_{23}}; \theta_{13;2})$$

over $\theta_{13;2}$ or when $\theta_{13;2}$ is univariate by inverting the empirical Kendall's τ based on the pseudo observations $(u_{k,1|2,\hat{\theta}_{12}}, u_{k,3|2,\hat{\theta}_{23}})$ for $k = 1, ..., n$.

This way of proceeding we extend now to higher dimensions.

Definition 7.3 (*Sequential estimation in simplified regular vines copulas*) We use the following notation:

- Let θ_e the copula parameter(s) corresponding to edge $e = (a_e, b_e; D_e)$ in a regular vine tree sequence in tree T_i.
- The copula parameters associated with the edges in tree T_i we denote by $\theta(T_i)$ and their estimates by $\hat{\theta}(T_i)$.
- Suppose all pair copula parameters identified in tree T_1 to tree T_{i-1} are already estimated. We denote this collection of parameter estimates by $\hat{\theta}(T_{1,...,i-1})$.

The sequential estimate of θ_e for edge $e = (a_e, b_e; D_e)$ in tree T_i is based on the *pseudo-observations*

$$u_{k,a_e|D_e,\hat{\theta}(T_{1,...,i-1})} := C_{a_e|D_e}(u_{k,a_e}|\boldsymbol{u}_{k,D_e}, \hat{\boldsymbol{\theta}}(T_{1,...,i-1}))$$

$$u_{k,b_e|D_e,\widehat{\boldsymbol{\theta}}(T_{1,...,i-1})} := C_{b_e|D_e}(u_{k,b_e}|\boldsymbol{u}_{k,D_e}, \widehat{\boldsymbol{\theta}}(T_{1,...,i-1}))$$

for $k = 1, ..n$. In particular, $\widehat{\boldsymbol{\theta}}_e$ is estimated by maximizing

$$\prod_{k=1}^{n} c_{a_e,b_e;D_e}(u_{k,a_e|D_e,\widehat{\boldsymbol{\theta}}(T_{1,...,i-1})}, u_{k,b_e|D_e,\widehat{\boldsymbol{\theta}}(T_{1,...,i-1})}; \theta_e)$$

or by inversion of the empirical Kendall's τ based on the pseudo-observations.

Ex 7.4 (`WINE3`: *Parameter estimates and log-likelihood*) We now present the sequential and joint maximum likelihood estimates and their log-likelihood for the three chosen pair copula constructions as defined in Example 4.3 in Table 7.1. From this, we see that the log-likelihoods are highest for the construction PCC1. This PCC has the highest empirical Kendall's τ values on the first tree. Further, the sequential estimation method using the inversion of the pairwise Kendall's τ estimates results in lower log-likelihoods in two cases compared to where the pair copula parameters are estimated sequentially using maximum likelihood. The improvement in the log-likelihood using joint maximum likelihood estimation is not so large compared to the sequential one using pairwise maximum likelihood.

7.3 Asymptotic Theory of Parametric Regular Vine Copula Estimators

We call a regular vine copula *parametric*, when all pair copulas are specified parametrically. Further, we assume that we have i.i.d. data available from a parametric regular vine copula. The associated parameters are collected in the parameter θ. In Sect. 7.2, we have developed methods to determine the sequential estimator and the joint MLE of θ. We first consider the case of maximum likelihood estimation.

The asymptotic theory for general maximum likelihood estimators is well developed (see for example, Lehmann and Casella 2006) and can be applied in this case. For the convenience of the reader, we recall the most relevant concepts. Under regularity conditions, one has for a p-dimensional parameter η with MLE $\hat{\eta}_n$ based on sample size n that

$$\sqrt{n}I(\eta)^{1/2}(\hat{\eta}_n - \eta) \to N_p(\mathbf{0}, I_p) \text{ as } n \to \infty, \tag{7.6}$$

where I_p is the $p \times p$-dimensional identity matrix and $I(\eta)$ is the expected Fisher-information matrix. Recall that for a log-likelihood of η given the random observation

Table 7.1 WINE3: Sequential (inversion of Kendall's τ, MLE) and joint ML estimates as well as associated Kendall's τ estimate together with the log-likelihoods

Edge	Family	Sequential $\hat{\theta}$	Inversion $\hat{\tau}$	Sequential $\hat{\theta}$	MLE $\hat{\tau}$	Joint $\hat{\theta}$	MLE $\hat{\tau}$
PCC1							
(acf, acc)	G	1.94	0.48	1.80	0.44	1.80	0.44
(acv, acc)	F	−4.57	−0.43	−4.44	−0.42	−4.51	−0.42
(acf, acv; acc)	N	0.20	0.13	0.17	0.11	0.17	0.11
log lik		807.90		816.83		816.92	
PCC2							
(acf, acv)	G270	−1.23	−0.19	−1.15	−0.13	−1.20	−0.16
(acc, acv)	F	−4.57	−0.43	−4.44	−0.42	−4.14	−0.40
(acf, acc; acv)	G	1.82	0.45	1.71	0.41	1.73	0.42
log lik		763.93		763.43		772.58	
PCC3							
(acf, acv)	G270	−1.23	−0.19	−1.18	−0.15	−1.17	−0.14
(acc, acf)	G	1.94	0.48	1.80	0.44	1.74	0.42
(acv, acc; acf)	F	−4.00	−0.39	−4.11	−0.40	−4.20	−0.40
log lik		770.27		784.05		785.72	

vector $X = (X_1^\top, \ldots, X_n^\top)^\top$ denoted by $\ell(\boldsymbol{\eta}|X)$ the Fisher-information matrix is defined as

$$I(\boldsymbol{\eta}) := -E_{\boldsymbol{\eta}}\left[\left(\frac{\partial^2}{\partial \eta_i \partial \eta_j}\ell(\boldsymbol{\eta}|X)\right)_{i,j=1,\ldots,p}\right]. \tag{7.7}$$

Since the expectation in (7.7) is often not analytically available, the Fisher-information matrix is then replaced by the observed one, which is defined as

$$I_n(\boldsymbol{\eta}) := -\left[\left(\frac{1}{n}\sum_{k=1}^{n}\frac{\partial^2}{\partial \eta_i \partial \eta_j}\ell(\boldsymbol{\eta}|x_k)\right)_{i,j=1,\ldots,p}\right]. \tag{7.8}$$

Here, $x_k = (x_{k1}, \ldots, x_{kd})^\top$ denotes the kth observation vector. This matrix $I_n(\boldsymbol{\eta})$ is by definition the negative Hessian matrix divided by n associated with the log-likelihood. An estimate of $I_n(\boldsymbol{\eta})$ can be obtained by replacing $\boldsymbol{\eta}$ by the maximum likelihood estimate $\hat{\boldsymbol{\eta}}_n$.

We apply now these general results to the regular vine copula likelihood specified in (7.1). Stöber and Schepsmeier (2013) develop a general algorithm to determine the observed Fisher-information matrix numerically associated with the copula parameter θ for an arbitrary regular vine copula. They use results of Schepsmeier and Stöber (2014) and the algorithm is implemented in the `VineCopula` package of Schepsmeier et al. (2018) in the function `RVineHessian`.

We now turn to the case of vine copula parameters, which have been sequentially estimated as defined in Sect. 7.2. This was first considered in Haff et al. (2013) and later improved and implemented in Stöber and Schepsmeier (2013) using a sandwich form of the asymptotic covariance matrix. For further details, see Stöber and Schepsmeier (2013).

One major application of the asymptotic result (7.6) is the construction of standard error estimates for the copula maximum likelihood parameter estimators. They are given by the square roots of the diagonal elements of the inverse of the observed information matrix $n \times I_n(\hat{\boldsymbol{\eta}}_n)$ based on (7.8). This is implemented in the function `RVineStdError` within the package `VineCopula`.

So far, we assumed that we have i.i.d. observations from a regular vine copula available. Now, we remark on the case, where only pseudo copula data is available resulting from estimated margins.

Remark 7.4 (Asymptotic theory for copula parameter estimators of regular vine based models) In applications, we do not have i.i.d. copula data but only an i.i.d. sample on the original scale available and so one has to consider also models for the marginal distribution. In particular, we can use a parametric or a nonparametric distribution for the margins. Joe and Xu (1996) follow a parametric approach, which they call *inference for margins (IFM) approach*. In contrast, Genest et al. (1995) utilize normalized ranks of the marginal observations, which can be considered as nonparametric estimates of the marginal distribution. We followed this approach in Example 3.4. Both approaches allow for asymptotic theory, however the resulting estimators of the marginal and copula parameters are less efficient than a joint approach, where the likelihood is maximized over marginal and copula parameters jointly. However, this is often numerically challenging and the loss of efficiency is often negligible (see Joe and Xu 1996 for examples).

Ex 7.5 (`WINE3`: *Standard error estimates for joint copula maximum likelihood estimates*) Estimated standard errors for the joint copula ML estimates are given in Table 7.2 for each of the pair copula constructions considered. Here, we assume that we have copula data available, i.e., we ignore the error made by estimating the margins using normalized ranks. The results show that all copula parameter estimates are statistically significant.

Table 7.2 WINE: Joint ML estimates together with estimated standard errors

Edge	Family	$\hat{\theta}$	$st\hat{d}er(\hat{\theta})$
PCC1			
(acf, acc)	G	1.80	0.04
(acv, acc)	F	−4.51	0.18
(acf, acv; acc)	N	0.17	0.02
PCC2			
(acf, acv)	G270	−1.20	0.02
(acc, acv)	F	−4.14	0.17
(acf, acc; acv)	G	1.73	0.04
PCC3			
(acf, acv)	G270	−1.17	0.02
(acc, acf)	G	1.74	0.04
(acv, acc; acf)	F	−4.20	0.18

7.4 Exercises

Exer 7.1
Likelihood for four-dimensional D- and C-vine copulas: Give for all terms in the likelihood expression (7.1) the dependence on the parameters for a four dimensional D- and C-vine copula, respectively.

Exer 7.2
WINE3: *Choice of pair copula families*: To investigate the choice of the pair copula family on the model fit, consider the data of Example 7.4. In this example, we have used careful exploratory analysis to choose the pair copula families. Refit now the three pair copula constructions with the following choices

- All pair copulas are bivariate Gaussian copulas.
- All pair copulas are bivariate Gumbel copulas.
- All pair copulas are bivariate Student's t copulas.

Compare the results to the ones presented in Example 7.4. What conclusions do you draw?

Exer 7.3
URAN3: *Pair copula models for the three dimensional* uranium *data*: Consider as in Exercise 4.1 the three-dimensional subset of the uranium data set contained in the R package copula with variables Cobalt (Co), Titanium (Ti) and Scandium

(Sc). Perform a similar analysis as in Example 7.4 using the choices made for the pair copula families in Example 4.1.

- For each PCC determine the sequential pair copula estimates using inversion of Kendall's τ and pairwise ML estimation, respectively.
- Determine the joint MLE for the three possible pair copula constructions.
- Which of the three pair copula constructions would you prefer?
- Determine standard error estimates of the joint ML estimates. What are the data assumptions and what conclusions can you draw given the assumptions?

Exer 7.4

ABALONE3 *Pair copula models for the three dimensional* abalone *data*: Perform the same analysis as in Exercise 7.3 for the three dimensional abalone data set considered in Exercise 1.6.

Selection of Regular Vine Copula Models

8

The full specification of a vine copula model requires the choice of a vine tree structure, copula families for each pair copula term and their corresponding parameters. In this chapter, we discuss different frequentist selection and estimation approaches. The three-layered definition of a regular vine copula leads to three fundamental estimation and selection tasks:

1. Estimation of copula parameters for a chosen vine tree structure and chosen pair copula families for each edge in the vine tree structure,
2. Selection of the parametric copula family for each pair copula term and estimation of the corresponding parameters for a chosen vine tree structure,
3. Selection and estimation of all three model components.

The first task we already have solved in Chap. 7. To fix ideas, we identify a full R-vine copula specification by the triplet $(\mathcal{V}, \mathcal{B}(\mathcal{V}), \Theta(\mathcal{B}(\mathcal{V})))$. Here, $\mathcal{V}$ denotes the vine tree structure, i.e., the building plan of the vine copula. The set $\mathcal{B}(\mathcal{V})$ collects all pair copula families associated with each edge of $\mathcal{V}$ and the set $\Theta(\mathcal{B}(\mathcal{V}))$ gives the associated pair copula parameters to each member of the set $\mathcal{B}(\mathcal{V})$.

Ex 8.1 (*Illustration of the general notation for a R-vine specification*) We now illustrate the general notation $(\mathcal{V}, \mathcal{B}(\mathcal{V}), \Theta(\mathcal{B}(\mathcal{V})))$ associated with the regular vine tree sequence $\mathcal{V}$ specified in Fig. 8.1. Here, $\mathcal{V} = \{T_i = (N_i, E_i), i = 1, \ldots, 4\}$ with node sets $N_1 = \{1, 2, 3, 4\}$, $N_2 = \{12, 23, 34, 35\}$, $N_3 = \{13; 2, 24; 3, 25; 3\}$, $N_4 = \{14; 23, 45; 23\}$ and edge sets $\{E_1 = N_2, E_2 = N_3, E_3 = N_4, E_4 = \{15; 234\}\}$. Suppose all pair copula families are Gaussian in the first tree T_1. For the second tree T_2 and the third tree T_3, we assume Gumbel and Clayton copulas, respectively. Then, the set of associ-

---

© Springer Nature Switzerland AG 2019
C. Czado, *Analyzing Dependent Data with Vine Copulas*, Lecture Notes in Statistics 222, https://doi.org/10.1007/978-3-030-13785-4_8

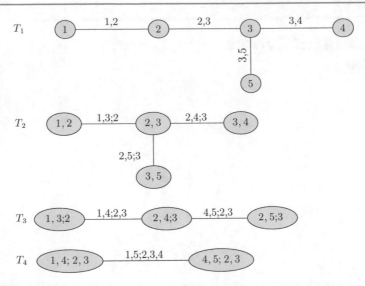

Fig. 8.1 R-vine trees: Regular vine tree sequence $\mathcal{V}$ on five elements for Example 8.1

ated pair copula families are $\mathcal{B}(\mathcal{V}) = \{$Gaussian for $e \in E_1$, Gumbel for $e \in E_2$, Clayton for $e \in E_3$ or $E_4\}$. As copula parameters, we select 0.5 for the Gaussian copulas, 1.5 for the Gumbel copulas and 1.2 for the Clayton copulas, then the set $\Theta(\mathcal{B}(\mathcal{V}))$ is specified as $\Theta(\mathcal{B}(\mathcal{V})) = \{0.5$ for $e \in E_1$, 1.5 for $e \in E_2$, 1.2 for $e \in E_3$ or $E_4\}$. So both sets $\mathcal{B}(\mathcal{V})$ and $\Theta(\mathcal{B}(\mathcal{V}))$ contain 10 elements.

8.1 Selection of a Parametric Copula Family for Each Pair Copula Term and Estimation of the Corresponding Parameters for a Given Vine Tree Structure

A d-dimensional regular vine copula with tree structure $\mathcal{V}$ is based on a set $\mathcal{B}(\mathcal{V})$ of $d(d-1)/2$ bivariate copula families and their corresponding parameters contained in $\Theta(\mathcal{B}(\mathcal{V}))$. The bivariate copula families can be chosen arbitrarily, e.g., from the popular classes of Archimedean, elliptical or extreme-value copulas. Assuming that an appropriate vine structure $\mathcal{V}$ has been specified, the question therefore is how to select adequate pair copula families $C_{a_e,b_e;D_e}$ and their parameters for given data $\boldsymbol{x}$ of sample size n in d dimensions. The (k, m)th data element of $\boldsymbol{x}$ is denoted by $x_{k,m}$ for $k = 1, \ldots, n$ and $m = 1, \ldots, d$.

Since we have to select bivariate parametric copula families in all trees, we use the Akaike information criterion (AIC) of Akaike (1998) to choose between bivariate copula families together with their parameters. This approach has been followed by Manner (2007) and Brechmann (2010). In particular, Brechmann (2010) compares the AIC-based bivariate copula selection to three alternative selection strategies: selection of the family with highest p-value of a copula goodness of fit test based on the Cramér-von Mises statistic (Genest et al. 2006, 2009), with smallest distance between empirical and modeled dependence characteristics (Kendall's τ, tail dependence), or with highest number of wins in pairwise comparisons of families using the test by Vuong (1989). In his large-scale Monte Carlo study, the AIC turned out to be the most reliable selection criterion for the pair copula family and its parameter(s). Therefore, we discuss now this *AIC based selection approach of the copula family and its parameter(s)* to all pair copula terms specified by the regular vine tree sequence $\mathcal{V}$.

More precisely, we first consider the selection of the pair copula families and their parameters in the first tree T_1 of $\mathcal{V}$. For an arbitrary edge $e = (a_e, b_e)$ in T_1, we use the copula data $u_{k,a_e} := F_{a_e}(x_{k,a_e})$ and $u_{k,b_e} := F_{b_e}(x_{k,b_e})$ for $k = 1, \ldots, n$. In particular, let $\mathcal{B}_e$ the set of possible parametric bivariate copula families for edge e. For each element C^B of $\mathcal{B}_e$, use the copula data $\boldsymbol{u}_e := \{u_{k,a_e}, u_{k,b_e}; k = 1, \ldots, n\}$ to fit this copula C^B with copula density c^B, i.e., find the maximum likelihood estimate of the associated parameter θ^B denoted by $\hat{\theta}^B$ as discussed in Sect. 3.10. In the next step, calculate the corresponding AIC value defined as

$$AIC(C^B, \hat{\theta}^B; \boldsymbol{u}_e) := -2 \sum_{k=1}^{n} \ln(c^B(u_{k,a_e}, u_{k,b_e}; \hat{\theta}^B)) + 2k^B, \qquad (8.1)$$

where k^B is the dimension of the parameter vector θ^B. Thus, we now select the copula family C^e with parameter θ^e for edge e, which minimizes $AIC(C^B, \hat{\theta}^B; \boldsymbol{u}_e)$ over the set $\mathcal{B}_e$. Note that this minimization always requires that all parameters are estimated for every pair copula family allowed in $\mathcal{B}_e$.

Since we most often fit only bivariate copula families with one- or two parameters, an alternative selection criterium such as the *Bayesian information criterion (BIC)* of Schwarz (1978) is not needed to induce model sparsity for the considered pair copula family selection.

For the copula family selection and parameter estimation of edges in trees T_i for $i > 1$, we use the sequential estimation approach developed in Sect. 7.2. For an edge $e = (a_e, b_e; D_e)$ in tree T_i for $i > 1$, we have pseudo-observations for the bivariate copula distribution of (U_{a_e}, U_{b_e}) given U_{D_e} available. These pseudo-observations were defined in Definition 7.3 and will be abbreviated as

$$\hat{u}_{k,a_e|D_e} := u_{k,a_e|D_e;\hat{\theta}(T_1,\ldots,T_{i-1})} \text{ and } \hat{u}_{k,b_e|D_e} := u_{k,b_e|D_e;\hat{\theta}(T_1,\ldots,T_{i-1})} \qquad (8.2)$$

for $k = 1, \ldots, n$. Now again consider the set of possible parametric bivariate copula families $\mathcal{B}_e$ for edge $e \in T_i$ for $i > 1$. The pseudo data $\boldsymbol{u}_e := \{\hat{u}_{k,a_e|D_e}, \hat{u}_{k,b_e|D_e}, k = 1, \ldots, n\}$ is used to estimate the copula parameter(s) θ^B by maximum likelihood for each element C^B of the set $\mathcal{B}_e$ by maximum likelihood. In the next step, determine the corresponding $AIC(C^B, \boldsymbol{\theta}^B, \boldsymbol{u}_e)$ similarly defined as in (8.1) using the pseudo data $\boldsymbol{u}_e$ for the edge $e = (a_e, b_e; D_e)$. Again we select that element from $\mathcal{B}_e$ which gives minimal AIC $AIC(C^B, \boldsymbol{\theta}^B, \boldsymbol{u}_e)$ value. This way of proceeding allows to select all pair copula families together with their parameter estimates in tree T_i. These choices are then used to define the pseudo-observations needed for the pair copula terms in tree T_{i+1}.

Ex 8.2 (Sequential selection of pair copula families) Consider again the five-dimensional regular vine tree sequence $\mathcal{V}$ of Fig. 8.1 with unknown pair copula families $\mathcal{B}(\mathcal{V})$ and their parameter set $\Theta(\mathcal{B}(\mathcal{V}))$. In the first tree, we need to select the copula families for $C_{1,2}$, $C_{2,3}$, $C_{3,4}$ and $C_{3,5}$ and estimate the corresponding parameters $\boldsymbol{\theta}_{1,2}, \boldsymbol{\theta}_{2,3}, \boldsymbol{\theta}_{3,4}$ and $\boldsymbol{\theta}_{3,5}$ using maximum likelihood based on copula observations $u_{k,j} := F_j(x_{kj})$, $k = 1, ..., n$ $j = 1, ..., 5$. For each possible element C^{B_e} in $\mathcal{B}_e$ estimate, the parameter $\boldsymbol{\theta}^{B_e}$ based on the copula data $\boldsymbol{u}_e$ for $e \in \{(1, 2), (2, 3), (3, 4), (3, 5)\}$ and determine the corresponding $AIC(C^{B_e}, \boldsymbol{\theta}^{B_e}; \boldsymbol{u}_e)$ values. Choose now for each of the four edges the copula family with the lowest AIC value.

Given these copula families and their estimated parameters for the first tree, we then have to select the families for the conditional pair copulas and their parameters in the second tree. In case of the copula $C_{1,3;2}$, we therefore again form the appropriate pseudo-observations $\hat{u}_{k,1|2} := C_{1|2}(u_{k,1}|u_{k,2}, \hat{\boldsymbol{\theta}}_{1,2})$ and $\hat{u}_{k,3|2} := C_{3|2}(u_{k,3}|u_{k,2}, \hat{\boldsymbol{\theta}}_{2,3})$, $k = 1, ..., n$,. Based on these, we estimate for each copula family allowed for this edge the associated copula family parameter. Select then the copula family $C_{1,3;2}$ together with its parameter estimate $\hat{\boldsymbol{\theta}}_{1,3;2}$ with the lowest AIC value.

Clearly, this sequential selection strategy accumulates uncertainty in the selection of the pair copula families and their associated parameter(s) over the tree levels. Therefore, the final model has to be carefully checked and compared to alternative models. For the latter, the tests for non-nested model comparison by Vuong (1989), Clarke (2007) may be used, see also Chap. 9.

8.2　Selection and Estimation of all Three Model Components of a Vine Copula

Since the pair copula family set $\mathcal{B}(\mathcal{V})$ and its associated parameter set $\theta(\mathcal{B}(\mathcal{V}))$ both depend on the vine tree structure $\mathcal{V}$, the identification of adequate trees is crucial to the model building process using vine copulas. As it was already the case for pair

Table 8.1 R-, C- and D-vines: Number of C-, D-, and R-vine tree structure for different values of d

d	Number of C- or D-vine structures	Number of R-vine structures
3	3	3
4	12	24
5	60	480
6	360	23040
7	2520	2580480
8	20160	660602880
9	181440	380507258880

copula selection in Sect. 8.1, it is again not feasible to simply try and fit all possible regular vine copula specification $(\mathcal{V}, \mathcal{B}(\mathcal{V}), \boldsymbol{\theta}(\mathcal{B}(\mathcal{V})))$ and then choose the "best" one. Recall that the number of possible regular vine tree sequences on d variables is $\frac{d!}{2} \times 2^{\binom{d-2}{2}}$ as shown by Morales-Nápoles (2011). This means that even if pair copula families and their parameters were known, the number of different models would still be enormous. This remains true, even when the selection is restricted to the subclasses of C- and D-vine structures, since there are still $d!/2$ different C- and D-vine tree structures in d dimensions, respectively (see Aas et al. (2009)). Table 8.1 gives the precise number of C- or D-vine and R-vine tree structures for dimensions $d = 3, \ldots, 9$. In particular in $d = 3$, all D-, C-, and R-vine structures coincide, while for $d = 4$, we have 12 C-vine and 12 D-vine structures. R-vine structures different than C- or D-vine structures only occur for dimensions $d > 4$. Table 8.1 again shows the enormous growth in the number of tree structures to choose from.

Two greedy construction strategies have been proposed in the literature: a top-down approach by Dißmann et al. (2013) and a bottom-up method by Kurowicka (2011). Both strategies proceed sequentially tree by tree and respect the proximity condition in each step. In this book we will restrict ourselves to the *top-down approach*. Since this approach is the most often used model selection algorithm for regular vine copulas and was first proposed by Dißmann et al. (2013), we also speak of the *Dißmann algorithm for regular vine copulas*.

8.3 The Dißmann Algorithm for Sequential Top-Down Selection of Vine Copulas

Selecting regular vine trees sequentially top-down means that we start with the selection of the first tree T_1 and continue tree by tree up to the last tree T_{d-1}. The first tree T_1 can be selected as an arbitrary spanning tree. Given that a tree T_m, $m \in \{1, ..., d-2\}$, has been selected, the next tree T_{m+1} is chosen respecting the proximity condition (see Remark 5.5). In other words, T_{m+1} can only be formed by the edges $\{a_e, b_e; D_e\}$, which satisfy the proximity condition.

Ex 8.3 (*Top-down tree selection*) Assume that we have selected the first tree T_1 as shown in Fig. 8.1. Then the question is which edges $\{a_e, b_e; D_e\}$ are eligible for the construction of the second tree T_2. According to the proximity condition, these are $\{1, 3; 2\}, \{2, 4; 3\}, \{2, 5; 3\}$, and $\{4, 5; 3\}$. Obviously, the last three pairs form a cycle and therefore only two of them can be selected for T_2. One of the three possibilities of choosing two edges is shown in Fig. 8.1.

To perform this iterative selection strategy, a criterion is needed to select a spanning tree among the set of eligible edges. Recall that a spanning tree simply denotes a tree on all nodes. Clearly, the *log-likelihood ℓ_m of the pair copulas in tree T_m* ($m = 1, \ldots, d - 1$) *of a regular vine copula* can be determined. For the tree T_m, we denote by $\mathcal{B}(T_m)$ the set of pair copula families for tree T_m and similarly, we use $\Theta(\mathcal{B}(T_m))$ for the set of corresponding parameters.

Definition 8.1 (*log-likelihood of tree T_m for a regular vine copula*) The log-likelihood of tree T_m ($m = 1, \ldots, d - 1$) with edge set E_m for a regular vine copula based on the copula data $\boldsymbol{u}$ is defined as

$$\ell_m\left(T_m, \mathcal{B}_{T_m}, \boldsymbol{\theta}_{T_m} | \boldsymbol{u}\right) = \sum_{k=1}^{n} \sum_{e \in E_m} \ln\left(c_{a_e, b_e; D_e}\left(C_{a_e | D_e}, C_{b_e | D_e}; \boldsymbol{\theta}_{a_e, b_e; D_e}\right)\right),$$

(8.3)

where we abbreviated $\mathcal{B}_{T_m} := \mathcal{B}(T_m)$ and $\boldsymbol{\theta}_{T_m} := \Theta(\mathcal{B}(T_m))$.

A straightforward solution therefore would be to choose the tree such that (8.3) is maximized after having selected pair copulas with high log-likelihood for each (conditional) pair $\{a_e, b_e; D_e\}$ satisfying the proximity condition. This solution however leads to highly over-parameterized models, since models with more parameters in which simpler models are nested will always give a higher likelihood. For instance, the Student's t copula always has a higher likelihood value than the Gaussian copula, since it is a special case of the Student's t copula as the degrees of freedom go to infinity.

Therefore we formulate the following algorithm in terms of a general weight ω assuming that we want to maximize it for each tree. The previously discussed strategy corresponds to choosing the pair copula log-likelihoods associated with an edge as edge weights.

Algorithm 8.2 (*Dißmann's algorithm*)

1: Calculate the weight $\omega_{i,j}$ for all possible index pairs $\{i, j\}$, $1 \leq i < j \leq n$.

2: Select the maximum spanning tree, i.e.

$$T_1 = \underset{T=(N,E) \text{ spanning tree}}{\operatorname{argmax}} \sum_{e=(a_e,b_e) \in E} \omega_{a_e,b_e}.$$

3: **for** each edge $e \in E_1$ **do**

4:     Select a copula C_{a_e,b_e} with estimated parameter(s) $\hat{\boldsymbol{\theta}}_{a_e,b_e}$ as discussed in Sect. 3.10.

5:     For $k = 1, \ldots, n$ generate the pseudo observations

$$C_{a_e|b_e}(u_{k,a_e}|u_{k,b_e}, \hat{\boldsymbol{\theta}}_{a_e,b_e}) \text{ and } C_{b_e|a_e}(u_{k,b_e}|u_{k,a_e}, \hat{\boldsymbol{\theta}}_{a_e,b_e})$$

as introduced in Definition 7.3.

6: **end for**

7: **for** $m = 2, \ldots, d - 1$ **do**

8:     Determine the weight $\omega_{a_e,b_e;D_e}$ for all edges $\{a_e, b_e; D_e\}$, that can be part of tree T_m. Denote the set of edges, which satisfy the proximity condition for tree T_m by $E_{P,m}$.

9:     Among these edges, select the maximum spanning tree, i.e.,

$$T_m = \underset{T=(N,E) \text{ spanning tree with } E \subset E_{P,m}}{\operatorname{argmax}} \sum_{e \in E} \omega_{a_e,b_e;D_e}.$$

10:     **for** each edge $e \in E_m$ **do**

11:         Select a pair copula $C_{a_e,b_e;D_e}$ with estimated parameter(s) $\hat{\boldsymbol{\theta}}_{a_e,b_e;D_e}$ as discussed in Sect. 8.1.

12:      For $k = 1, ..., n$ generate the pseudo observations

$C_{a_e|b_e\cup D_e}(u_{k,a_e}|u_{k,b_e}, u_{k,D_e}, \hat{\theta}_{a_e,b_e|D_e})$ and $C_{b_e|a_e\cup D_e}(u_{k,b_e}|u_{k,a_e}, u_{k,D_e}, \hat{\theta}_{a_e,b_e|D_e})$

using Definition 7.3.

13:    **end for**

14: **end for**
15: **return** the sequential model estimate $(\hat{\mathcal{V}}, \hat{\mathcal{B}}, \hat{\theta})$.

Clearly, this algorithm only makes a locally optimal selection in each step, since the impact on previous and subsequent trees is ignored. The strategy is however reasonable with regard to statistical modeling with regular vine copulas. Important dependencies as measured by the weights are modeled first. The maximum spanning tree in lines 2 and 10 can be found, e.g., using the classical algorithms by Prim or Kruskal (see for example Sect. 23.2 of Cormen et al. (2009)). Later Gruber and Czado (2015) have shown in six-dimensional simulation scenarios, that the greedy Dißmann algorithm gets about at least 75% of the true log-likelihood, which indicates a good performance. Possible choices for the weight ω are, for example,

- the absolute empirical Kendall's τ as proposed by Dißmann et al. (2013), Czado et al. (2012);
- the AIC of each pair copula as introduced in Sect. 8.1;
- the (negative) estimated degrees of freedom of Student's t pair copulas as proposed by Mendes et al. (2010);
- the p-value of a copula goodness of fit test and variants as proposed by Czado et al. (2013).

Remark 8.3 (*Sequential selection of C- and D-vine copulas*) Algorithm 8.2 can easily be modified to select C- or D-vine copulas instead of general R-vine copulas. For C-vine structures the root node in each tree can simply be identified as the node with the maximal sum of weights compared to all other nodes (see Czado et al. (2012)). In the case of D-vines structures only the order of variables in the first tree has to be chosen. Since D-vine trees are Hamiltonian paths, a maximum Hamiltonian path has to be selected. This problem is equivalent to a traveling salesman problem as discussed by Brechmann (2010). As an NP-hard problem, there is no known efficient algorithm to find a solution. In practise we have utilized the TSP package of Hahsler and Hornik (2017) in R, which proposes several solutions for the order of the D-vine.

Fig. 8.2 `WINE7`: Complete graph of all pairs of variables of the first tree

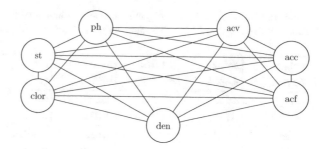

Ex 8.4 (`WINE7`: *Illustration of the Dißmann Algorithm*) To illustrate the model selection algorithm discussed in this section, we again study the extended data set of the chemical components of red wines considered in Exercises 1.7 and 3.3.

The original data is rank transformed for each margin and the corresponding empirical pairwise copula data and normalized contour plots are displayed in Fig. 3.15. Many of the normalized contour plots (lower triangular panels) show non-elliptical shapes, thus indicating the need to allow for non-Gaussian dependence. Pairwise empirical Kendall's τ, Pearson correlations and Spearman's correlations are given in Table 2.3. We see that strong positive and negative dependence among the variables are present.

We illustrate the Dißmann Algorithm 8.2 using absolute values of the pairwise empirical Kendall's τ values as weights. Table 8.2 shows the ordered weights needed to determine the first vine tree T_1.

Figure 8.2 shows the complete graph of all pairs of variables, while Fig. 8.3 displays the selected edges for T_1. It also illustrates the restriction to a tree in the selection of the edges. Finally Fig. 8.4 shows the selected first tree. This completes the selection of the first tree T_1. After this, we select and estimate for each edge in tree T_1 the pair copula family and the associated parameter using the AIC criterion defined in (8.1), respectively.

Now, we turn to the selection of tree T_2. For this, we generate the pseudo copula data for tree T_2 using (8.2). More precisely, the pseudo data $\hat{u}_{k,\mathrm{den;acf}}$, $\hat{u}_{k,\mathrm{ph;acf}}$, $\hat{u}_{k,\mathrm{acf;acc}}$, $\hat{u}_{k,\mathrm{acv;acc}}$, $\hat{u}_{k,\mathrm{acc;acf}}$, $\hat{u}_{k,\mathrm{st;clor}}$, $\hat{u}_{k,\mathrm{den;clor}}$, $\hat{u}_{k,\mathrm{acf;den}}$, and $\hat{u}_{k,\mathrm{clor;den}}$ for $k = 1, \ldots, n$ is needed. It is used to estimate the Kendall's τ associated with the potential pair copula terms in tree T_2. The resulting Kendall's τ estimates are given in Table 8.3. All possible edges according to the proximity condition are shown in Fig. 8.5. The edge choices which are available for tree T_2 are illustrated in Fig. 8.6. The final choices are then indicated in bold in Fig. 8.7. Since tree T_2 is a Hamilitonian path as in a D-vine tree structure, the tree choices for trees T_3 until T_6 are determined.

Table 8.2 WINE7: Ordered weights (pairwise absolute empirical Kendall's τ values of copula data) for all pairs of variables. The edges chosen according to Algorithm 8.2 are highlighted in bold

acf,ph	acf,acc	acf,den	acv,acc	acc,ph	clor,den	acc,den
0.5278	**0.4843**	**0.4575**	**0.4284**	0.3898	**0.2879**	0.2457
den,ph	acf,acv	acf,clor	clor,ph	acv,ph	acv,clor	clor,st
0.2172	0.1852	0.1760	0.1627	0.1587	0.1090	**0.0916**
st,den	acc,clor	acv,st	acf,st	acv,den	acc,st	st,ph
0.0877	0.0767	0.0637	0.0569	0.0159	0.0116	0.0068

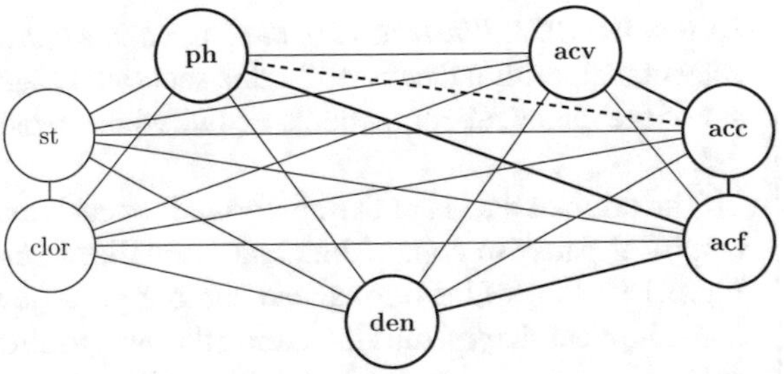

Fig. 8.3 WINE7: Illustration of stepwise selection of the highest weights. Selected edges are shown as solid black lines. Selecting the dashed edge would result in a cycle within the graph and is therefore not allowed to be selected

Fig. 8.4 WINE7: First tree graph shown with selected edges in bold

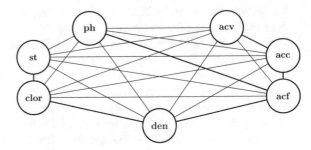

Fig. 8.5 WINE7: Graph showing all pairs of variables for tree T_2 and edges allowed by the proximity condition

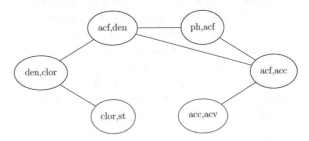

Fig. 8.6 WINE7: Illustration of stepwise selection of the highest weights for Tree T_2. Selected edges are illustrated as solid bold lines. Selecting the dashed edge would result in a cycle within the graph and is not allowed to be selected

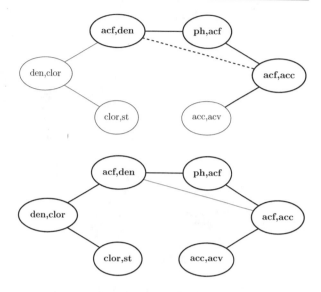

Fig. 8.7 WINE7: Second tree graph with selected edges highlighted in bold

Table 8.3 WINE7: Ordered weights (pairwise absolute empirical Kendall's τ values of pseudo data based on the copula fits of tree T_1) for all pairs of pseudo variables. The edges chosen according to the Dißmann Algorithm 8.2 are highlighted in bold

den,ph;acf	acf,acv;acc	acc,ph;acf	acc,den;acf	st,den;clor	acf,clor;den
0.1894	**0.1283**	**0.0978**	0.0831	**0.0484**	**0.0050**

Ex 8.5 (WINE7 *Fitted R-vine, C-, and D-vine copulas*)   We again consider the rank transformed data of Example 8.4. The function RVine StructureSelect implements the Algorithm 8.2 proposed by Dißmann et al. (2013). For this data set, we allowed for the Gaussian (N), Frank (F), Joe (J), and Gumbel (J) copula and its rotations of the Joe (SJ, J90 or J270) and of the Gumbel (SG, G90 or G270) as pair copula families. As model selection criteria for the pair copula families the AIC criterion (8.1) is used.

We fitted R-vine, C-, and D-vine copulas sequentially with the pair copula families given above as well a *Gaussian R-vine copula*, where all pair copulas are bivariate Gaussian copula. The TSP package gave den-acf-ph-acc-acv-st-clor as D-vine order.

The resulting tree sequence plots are given in Figs. 8.8, 8.9 and 8.10 for the R-vine, C-vine, and D-vine, respectively. We see that the fitted R-vine is close to a D-vine, while the C-vine is different.

When the dependence strength in a pair copula is weak, we might conduct an independence test between the pair of variables based on Kendall's τ as given in (2.8). If the tests fails to reject the null hypothesis of independence at the 5%

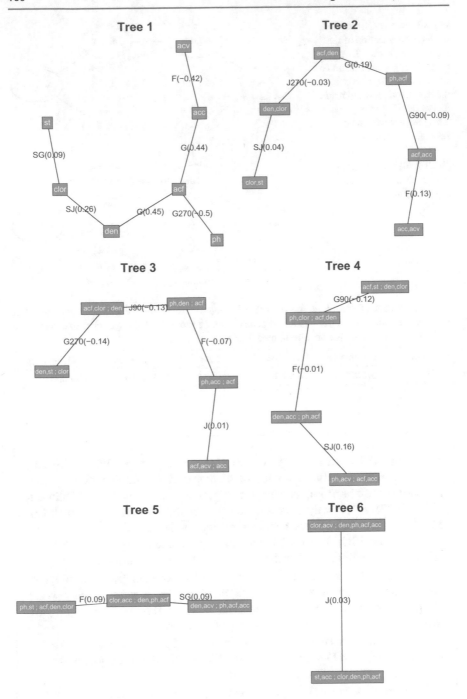

Fig. 8.8 WINE7: Selected R-vine tree plots based on the Dißmann Algorithm 8.2

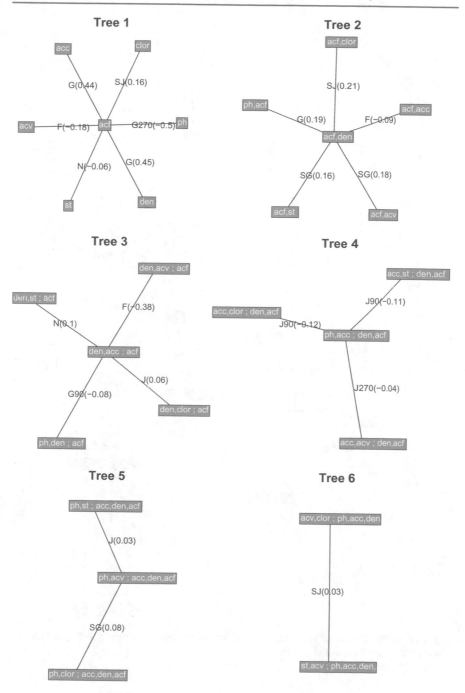

Fig. 8.9 WINE7: Selected C-vine tree plots based on Dißmann Algorithm 8.2

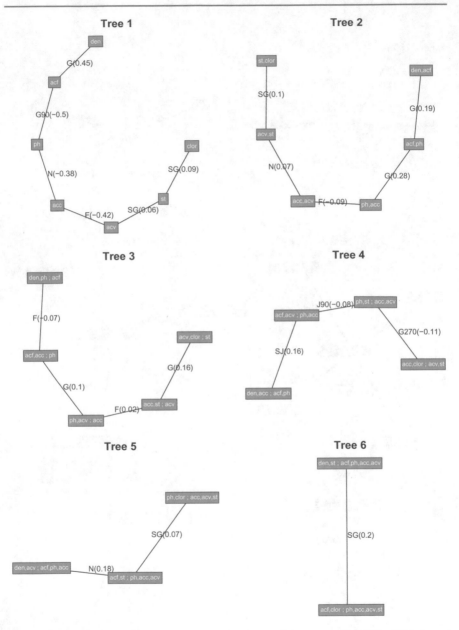

Fig. 8.10 WINE7 : Selected D-vine tree plots based on one solution found by TSP of Hahsler and Hornik (2017) for the order of the variables

Table 8.4 WINE7: Chosen pair copula families for different vine copula models. Models with *ind* at the ending denote a selection with an independence test performed for each pair copula in the Dißmann Algorithm 8.2

	N	F	J	G	SJ	J90	J270	SG	G90	G270	I
R-vine	0	5	2	3	3	1	1	2	2	2	0
R-vine-ind	0	4	1	3	3	1	0	2	2	2	3
C-vine	2	3	2	3	3	2.	1	3	1	1	0
C-vine-ind	2	3	1	4	2	1	1	2	2	1	2
D-vine	3	4	0	5	1	1	0	5	1	1	0
D-vine-ind	3	3	0	5	1	1	0	5	1	1	1

significance level, then we replace the pair copula by the independence copula. We can also apply the Dißmann Algorithm 8.2 with choosing the independence copula for the pair copula term in this case. This approach identified at most three associated pair copulas, which can be modeled by the independence pair copula for this data set as can be seen from Table 8.4. In Table 8.4 the occurrence of the pair copula families for the different models are recorded. From this table we see that a Gaussian R-vine is not selected for this data, since many of the pair copula terms are not selected as a Gaussian copula.

Finally we report the range of the estimated (conditional) Kendall's τ values corresponding to all pair copulas of all models without the independence test. Their values are given in Table 8.5 for the R-vine, Gaussian R-vine, C-vine, and D-vine, respectively. Copula parameters are jointly estimated using maximum likelihood. This table shows that the selected vine models have decreasing dependence strength in the pair copulas as the level j of tree T_j increases. Here, we assess the dependence strength by the fitted conditional Kendall's τ values.

8.4 Exercises

Exer 8.1

(ABALONE6: *Dependence models for the six dimensional* abalone *data set*: We now allow additional variables of the abalone data set contained in the R package PivotalR from Pivotal Inc. (2017). Here, we consider the female abalone shells only and use the variables length, diameter, whole, shucked viscera and shell. We transform them to the copula scale by using empirical distribution functions.

Table 8.5 `WINE7`: Range of estimated (conditional) Kendall's tau values per tree for MLE estimated R-vine model (top left), Gaussian R-vine model (top right), C-vine model (bottom left), and D-vine model (bottom right)

R-vine	Min	Max
Tree 1	−0.50	0.46
Tree 2	−0.09	0.19
Tree 3	−0.13	0.01
Tree 4	−0.13	0.16
Tree 5	0.09	0.09
Tree 6	0.03	0.03
Gaussian		
Tree 1	−0.51	0.46
Tree 2	−0.11	0.18
Tree 3	−0.15	−0.02
Tree 4	−0.14	0.18
Tree 5	0.07	0.11
Tree 6	0.04	0.04
C-vine		
Tree 1	−0.49	0.46
Tree 2	−0.08	0.20
Tree 3	−0.38	0.10
Tree 4	−0.12	−0.03
Tree 5	0.03	0.08
Tree 6	0.04	0.04
D-vine		
Tree 1	−0.49	0.46
Tree 2	−0.09	0.29
Tree 3	−0.13	0.09
Tree 4	−0.13	0.16
Tree 5	0.08	0.09
Tree 6	0.03	0.03

- Use the Dißmann Algorithm 8.2 to select an R-vine copula, where all implemented pair copulas of the R package `VineCopula` of Schepsmeier et al. (2018) are used with no zero Kendall's τ test. As the estimation method uses the sequential approach.
- Perform the same analysis as above for selecting a C- and a D-vine copula.
- Illustrate the associated vine tree structure of the vine copula models studied.
- Investigate if a zero Kendall's τ test can induce more parsimonious vine copula models.

- Investigate which pair copula families have been chosen for each vine copula class.
- Study the fitted strength of dependence, the occurrence of tail dependence, the occurrence of asymmetries for the selected pair copulas as the tree level increases.
- Determine joint maximum likelihood estimates of the copula parameters for all studied models and compare them to the corresponding sequential ones.

Exer 8.2

(URANIUM7: *Dependence models for the seven dimensional* uranium *data set*: Consider now all variables of the uranium data set contained in the R package copula of Hofert et al. (2017). Perform a similar analysis as in Exercise 8.1.

Comparing Regular Vine Copula Models

9

In this chapter, we want to compare the fit of two or more regular vine copula specifications for a given copula data set. For each considered model, we have to specify

- the vine tree structure $\mathcal{V}$ and
- the set of pair copula families $\mathcal{B}(\mathcal{V}) = \{B_e | i = 1, \ldots, d-1; e \in E_i\}$.

Recall that the set of corresponding copula parameters are denoted by $\theta(\mathcal{B}(\mathcal{V}))$. For a random sample $u = (u_1^\top, \ldots, u_n^\top)^\top$ of size n from this model specification with the kth observation $u_k \in [0, 1]^d$, the associated *likelihood* for the parameters $\theta(\mathcal{B}(\mathcal{V}))$ in a regular vine model with vine structure $\mathcal{V}$ and pair copula family set $\mathcal{B}(\mathcal{V})$ can be expressed as

$$\ell(\theta(\mathcal{B}(\mathcal{V})); u) = \prod_{k=1}^{n} \ell_k(\theta(\mathcal{B}(\mathcal{V})); u_k), \tag{9.1}$$

where

$$\ell_k(\theta(\mathcal{B}(\mathcal{V})); u_k) := \prod_{i=1}^{d-1} \prod_{e \in E_i} c_{a_e, b_e; D_e}(C_{a_e|D_e}(u_{k,a_e}|u_{k,D_e}), C_{b_e|D_e}(u_{k,b_e}|u_{k,D_e}))$$

denotes the likelihood contribution of the kth observation u_k (compare to Eq. (7.1)). This allows us to use classical model comparison criteria for likelihood-based statistical models.

© Springer Nature Switzerland AG 2019
C. Czado, *Analyzing Dependent Data with Vine Copulas*, Lecture Notes
in Statistics 222, https://doi.org/10.1007/978-3-030-13785-4_9

9.1   Akaike and Bayesian Information Criteria for Regular Vine Copulas

Classical statistical model comparison criteria are the Akaike information criterion (AIC) of Akaike (1973) and the Bayesian information criterion (BIC) of Schwarz (1978). We give now the general form of *AIC* and *BIC for arbitrary regular vine copulas*. Recall that we already encountered AIC, when selecting a pair copula term within a regular vine copula (see Eq. (8.1)).

Definition 9.1 (*Akaike and Bayesian information criteria for regular vine copula models*) For an random copula sample u of size n from a regular vine copula with triplet $(\mathcal{V}, \mathcal{B}(\mathcal{V}), \theta(\mathcal{B}(\mathcal{V})))$, the Akaike information criterion for the complete R-vine copula specification is defined as

$$AIC_{RV} = -2 \sum_{k=1}^{n} \ln(\ell_k(\hat{\theta}(\mathcal{B}(\mathcal{V})); u_k) + 2K, \qquad (9.2)$$

while corresponding Bayesian information criterion is

$$BIC_{RV} = -2 \sum_{k=1}^{n} \ln(\ell_k(\hat{\theta}(\mathcal{B}(\mathcal{V})); u_k) + \ln(n)K. \qquad (9.3)$$

Here K is number of model parameters, i.e., the length of the vector $\theta(\mathcal{B}(\mathcal{V}))$. Further $\hat{\theta}(\mathcal{B}(\mathcal{V}))$ denotes an estimate of $\theta(\mathcal{B}(\mathcal{V}))$.

Ex 9.1 (`WINE3`: *Model selection based on Akaike and Bayesian information criteria*) The AIC_{RV} and BIC_{RV} values of the three PCC's copula models discussed in Example 7.4 are given in Table 9.1. For this data set we have two sequential parameter estimation methods available, one using the inversion of Kendall's τ for each pair copula parameter and the other one based on maximum likelihood estimation of each pair copula parameter. The third estimation method is standard maximum likelihood estimation of all parameters jointly. The results in Table 9.1 show that the sequential pairwise maximum likelihood estimation is closer to the joint maximum likelihood estimation compared to the sequential pairwise inversion of Kendall's τ estimation for the PCC1 and PCC3 model. This holds for both AIC_{RV} and BIC_{RV} values. According to both AIC_{RV} and BIC_{RV} the PCC1 copula model specification has the lowest values, thus it is the preferred vine copula model for this data set. For the PCC2 specification the sequential estimation with pairwise inversion of Kendall's τ estimation performs better than the sequential estimation with pairwise maxi-

Table 9.1 WINE3: AIC_{RV} and BIC_{RV} for PCC1 (D-vine order acf-acc-acv), PCC2 (D-vine order acf-acv-acc) and PCC3 (D-vine order acv-acf-acc) copula models using three parameter estimation methods (sequential pairwise inversion of Kendall's τ, pairwise and joint maximum likelihood)

PCC copula	Sequential estimation Pairwise τ inversion	Sequential estimation Pairwise ML	Joint MLE
AIC_{RV}			
PCC1: acf-acc-acv	−1609.8	−1627.7	−1627.8
PCC2: acf-acv-acc	−1521.9	−1520.9	−1539.2
PCC3: acv-acf-acc	−1534.5	−1562.1	−1565.4
BIC_{RV}			
PCC1: acf-acc-acv	−1593.5	−1611.5	−1611.7
PCC2: acf-acv-acc	−1505.7	−1504.7	−1523.0
PCC3: acv-acf-acc	−1518.4	−1546.0	−1549.3

mum likelihood. However, in this case, the joint maximum likelihood estimates yield a higher improvement of the log-likelihood than the sequential estimation methods for the PCC2. For the other two specifications PCC1 and PCC2 the sequential estimation using pairwise maximum likelihood are very close to the joint maximum likelihood estimation results.

Ex 9.2 (WINE7: Model selection based on AIC and BIC) We now continue our analysis of the extended wine data considered in Exercises 1.7 and 3.3. We use the fitted vine copula models in Example 8.5 for this data set. In addition to the sequential parameter estimates the corresponding joint maximum likelihood estimates were also determined, showing only a small increase of the log-likelihood fit over the sequentially fitted ones. Table 9.2 summarizes the fit of the studied vine copula models.

The overall AIC_{RV} criterion identifies the regular vine without the reduction of using independence copulas as the best fitting model, while the BIC_{RV} selects the regular vine with the reduction of using independence pair copulas identified by the independence test for the null hypothesis $\tau = 0$ for each pair copula term. This is a result of the fact that the BIC criterion, in general, prefers more parsimonious models.

We see that regular vine copulas provide a better fit than the subclasses of C- and D-vine copulas. Comparing C-vines and D-vines specifications we note, that C-vine copulas are preferred over D-vine copulas, when the model is not reduced by applying the independence test. If the reduced model approach is

Table 9.2 WINE7: Estimated log-likelihood, number of parameters (# par), AIC_{RV} and BIC_{RV} for all fitted models and estimation methods (seq = sequential estimation, mle = maximum likelihood, ind = independence test used to allow for independence pair copula family)

Model	Method	Log-likelihood	# par	AIC_{RV}	BIC_{RV}
R-vine	seq	2525.32	21	−5008.64	−4895.72
R-vine	mle	2527.30	21	−5012.59	−4899.67
R-vine	ind-seq	2519.48	18	−5002.96	−4906.17
R-vine	ind-mle	2521.17	18	−5006.34	−4909.55
Gauss	seq	2271.77	21	−4501.55	−4388.63
Gauss	mle	2271.80	21	−4501.59	−4388.67
Gauss	ind-seq	2266.98	18	−4497.95	−4401.16
Gauss	ind-mle	2267.15	18	−4498.30	−4401.51
C-vine	seq	2489.41	21	−4936.81	−4823.89
C-vine	mle	2496.41	21	−4950.83	−4837.91
C-vine	ind-seq	2446.52	19	−4855.05	−4752.88
C-vine	ind-mle	2454.46	19	−4870.91	−4768.75
D-vine	seq	2445.21	21	−4848.31	−4735.37
D-vine	mle	2452.74	21	−4863.48	−4750.58
D-vine	ind-seq	2443.60	20	−4846.89	−4739.42
D-vine	ind-mle	2551.23	20	−4862.40	−4754.80

used, this is reversed, since the decrease in the log-likelihood of the reduced model is larger for C-vine than for D-vine copulas.

Finally we compare to the Gaussian vine copula fits, where the pair copula family is fixed to be a bivariate Gaussian copula. Note that all Gaussian vine copula fits correspond to a d-dimensional Gaussian copula with however possibly different correlation parameter estimates. Since the Gaussian AIC_{RV} and BIC_{RV} values are much lower than the corresponding regular, D- and C-vine copula fits, we conclude that the dependence structure present in this data set is not a Gaussian one.

Remark 9.2 (AIC and BIC for non-nested models) Often the AIC and BIC criteria are only used when comparing nested models. We speak of *nested models*, when we have a "full" model, that is specified in terms of certain parameters, while the "reduced" model is a special case of this "full" model and parametrized with a subset of these parameters. However, Ripley (2008, pp. 34–35) states that AIC comparisons are also feasible for non-nested models but at the expense of an increased variability of the estimated AIC difference for pairs of non-nested models.

9.2 Kullback–Leibler Criterion

Since we are interested in comparing non-nested vine copula models, we introduce now an alternative method for arbitrary parametric statistical models proposed by Vuong (1989). It is a statistical test and thus allows to assign a significance level between two different statistical models, while the AIC and BIC criteria only allow for a preference order among the studied models. To discuss the Vuong test we first need to introduce the *Kullback–Leibler criterion (KLIC)* of Kullback and Leibler (1951).

The Kullback–Leibler criterion (KLIC) measures the distance between the true but unknown density h_0 and a specified, approximating parametric density $f(\cdot|\theta)$ with parameter θ. Note that the parameter θ does need to coincide with the parameter of the true density h_0, if h_0 is a parametric density. We follow here the exposition of Vuong (1989).

Definition 9.3 (*Kullback–Leibler criterion (KLIC) for comparing statistical models*) The Kullback–Leibler criterion (KLIC) between a true density $h_0(\cdot)$ of a random vector X and the approximating density $f(\cdot|\theta)$ is defined as

$$KLIC(h_0, f, \theta) := \int \ln\left[\frac{h_0(x)}{f(x|\theta)}\right] h_0(x)dx = E_0[\ln h_0(X)] - E_0[\ln f(X|\theta)], \quad (9.4)$$

where E_0 denotes the expectation with respect to the true distribution with density h_0.

The optimal choice of θ among all allowed parameter vectors with regard to the KLIC criterion given in (9.4) is the one which maximizes $E_0[\ln f(X|\theta)]$. Given i.i.d. observations $x_k, k = 1, \ldots, n$ from the true density h_0, the term $E_0[\ln f(X|\theta)]$ can be estimated by

$$\frac{1}{n} \sum_{k=1}^{n} \ln f(x_k|\theta).$$

The associated maximum likelihood estimate of θ based on a sample of size n, denoted by $\hat{\theta}_n$, maximizes this term. Thus the minimum of $KLIC(h_0, f, \theta)$ over all parameter values is estimated by $KLIC(h_0, f, \hat{\theta}_n)$.

Remark 9.4 (*KLIC and Kullback–Leibler divergence*) The Kullback–Leibler criterion is the special case of the *Kullback–Leibler divergence* between two densities h_0

and h_1, defined as

$$KLIC(h_0, h_1) = E_0(\ln\left[\frac{h_0(X)}{h_1(X)}\right])$$

for $h_1x) = f(x|\theta)$.

9.3 Vuong Test for Comparing Different Regular Vine Copula Models

Vuong (1989) was interested in assessing, whether two competing model classes $\mathcal{M}_j$ specified by parametric densities $\{f_j(\cdot|\theta) \text{ for } \theta \in \Theta_j\}$ for $j = 1, 2$ are providing equivalent fits for a given data set or whether one model class is to be preferred over the other. Let θ_j be the value of $\theta \in \Theta_j$, which maximizes $E_0(\ln f_j(X|\theta))$ over $\theta \in \Theta_j$ for $j = 1, 2$. In particular, he investigated the cases, where the two model classes are *strictly non-nested, overlapping or nested*. Strictly non-nested models do not share parameters, while overlapping models share some but not all parameters. Recall that two models are called nested, if one model is a special case of the other, thus the parameters of the smaller model are contained in the set of parameters for the larger model.

More precisely, Vuong (1989) utilized the KLIC defined in (9.4) and constructed an asymptotic likelihood ratio test for the null hypothesis of equivalence between the two model classes $\mathcal{M}_1$ and $\mathcal{M}_2$:

$$H_0 : KLIC(h_0, f_1, \theta_1) = KLIC(h_0, f_2, \theta_2).$$

Using (9.4) this null hypothesis can be equivalently expressed as

$$H_0 : E_0\left[\ln f_1(X|\theta_1)\right] = E_0\left[\ln f_2(X|\theta_2)\right]. \tag{9.5}$$

Obviously, when

$$H_{m1} : E_0[\ln f_1(X|\theta_1)] > E_0[\ln f_2(X|\theta_2)],$$

holds model $\mathcal{M}_1$ is better than $\mathcal{M}_2$. The corresponding reverse hypothesis to H_{m1} we call H_{m2}. In this case model $\mathcal{M}_2$ is better than model $\mathcal{M}_1$. However, the question arises, whether one of the models is *significantly* better than the other. An obvious test statistics for the null hypothesis in (9.5) is given by the likelihood ratio.

Definition 9.5 (*Likelihood ratio statistic for two approximating statistical models*) For i.i.d. observations x_k, $k = 1, ..., n$, from the true density h_0 with maximum likelihood estimates $\hat{\theta}_j$ associated with statistical model classes $\mathcal{M}_j$ specified by the densities $f_j(\cdot|\theta_j)$ for $j = 1, 2$ define the kth observed log-likelihood ratio contribution

$$m_k(\boldsymbol{x}_k) := \ln\left[\frac{f_1(\boldsymbol{x}_k|\hat{\boldsymbol{\theta}}_1)}{f_2(\boldsymbol{x}_k|\hat{\boldsymbol{\theta}}_2)}\right], \; k = 1, ..., n.$$

The observed likelihood ratio statistic for the two approximating model $\mathcal{M}_1$ and model $\mathcal{M}_2$ is defined by

$$LR_n(\hat{\boldsymbol{\theta}}_1, \hat{\boldsymbol{\theta}}_2)(\boldsymbol{x}_1, \ldots, \boldsymbol{x}_n) := \sum_{k=1}^{n} m_k(\boldsymbol{x}_k). \tag{9.6}$$

Further, let $m_k(\boldsymbol{X}_k)$ and $LR_n(\hat{\boldsymbol{\theta}}_1, \hat{\boldsymbol{\theta}}_2)(\boldsymbol{X}_1, \ldots, \boldsymbol{X}_n)$ the associated random quantities, when $\boldsymbol{X}_k \sim h_0$ i.i.d.

Under the true distribution h_0 and appropriate regularity conditions on the two model classes, the normalized likelihood ratio $\frac{1}{n}LR_n(\hat{\boldsymbol{\theta}}_1, \hat{\boldsymbol{\theta}}_2)(\boldsymbol{X}_1, \ldots, \boldsymbol{X}_n)$ is a random variable satisfying the following law of large numbers.

Theorem 9.6 (Law of large numbers for the LR statistics) *Under the regularity conditions in Lemma 3.1 of Vuong (1989), it follows that*

$$\frac{1}{n}LR_n(\hat{\boldsymbol{\theta}}_1, \hat{\boldsymbol{\theta}}_2)(\boldsymbol{X}_1, \ldots, \boldsymbol{X}_n) = \frac{1}{n}\sum_{k=1}^{n} m_k(\boldsymbol{X}_k) \xrightarrow[n\to\infty]{a.s.} E_0\left(\ln\frac{f_1(\boldsymbol{X}|\boldsymbol{\theta}_1)}{f_2(\boldsymbol{X}|\boldsymbol{\theta}_2)}\right).$$

Vuong (1989) now considers the sample variance of the likelihood ratio statistics $LR_n(\hat{\boldsymbol{\theta}}_1, \hat{\boldsymbol{\theta}}_2)(\boldsymbol{X}_1, \ldots, \boldsymbol{X}_n)$

$$\hat{\omega}(\boldsymbol{x}_1, \ldots, \boldsymbol{x}_n)^2 := \frac{1}{n}\sum_{k=1}^{n}(m_k(\boldsymbol{x}_k) - \bar{m}(\boldsymbol{x}_1, \ldots, \boldsymbol{x}_n))^2,$$

where $\bar{m}(\boldsymbol{x}_1, \ldots, \boldsymbol{x}_n) = n^{-1}\sum_{k=1}^{n} m_k(\boldsymbol{x}_k)$ and obtains the asymptotic distribution of $LR_n(\hat{\boldsymbol{\theta}}_1, \hat{\boldsymbol{\theta}}_2)(\boldsymbol{X}_1, \ldots, \boldsymbol{X}_n)$.

Theorem 9.7 (Asymptotic normality of the likelihood ratio statistics for two approximating strictly non-nested models) *Under regularity conditions of Theorem 5.1 in Vuong (1989) the following asymptotic normality under the* *null hypothesis* (9.5) *holds*

$$\nu(X_1, \ldots, X_n) := \frac{LR_n(\hat{\theta}_1, \hat{\theta}_2)(X_1, \ldots, X_n)}{\sqrt{n\hat{\omega}(X_1, \ldots, X_n)^2}} \xrightarrow[n \to \infty]{\mathcal{D}} N(0, 1). \qquad (9.7)$$

This readily yields an asymptotic test called the *Vuong test* for model selection between strictly non-nested models. Since the Vuong test is a statistical test, it also assigns a significance level to its decision or does not make a decision at all at the prespecified level.

Definition 9.8 (*Asymptotic α level Vuong test for strictly non-nested models*) An asymptotic α level test of

$$H_0 : E_0\left[\ln f_1(X|\boldsymbol{\theta}_1)\right] = E_0\left[\ln f_2(X|\boldsymbol{\theta}_2)\right] \text{ against } H_1 : \text{ not } H_0, \qquad (9.8)$$

is given by:

$$\text{Reject } H_0 \text{ if and only if } |\nu(\boldsymbol{x}_1, \ldots, \boldsymbol{x}_n)| > \Phi^{-1}\left(1 - \tfrac{\alpha}{2}\right).$$

In particular,

- if $\nu(\boldsymbol{x}_1, \ldots, \boldsymbol{x}_n) > \Phi^{-1}\left(1 - \tfrac{\alpha}{2}\right)$, we prefer model class $\mathcal{M}_1$ to model class $\mathcal{M}_2$, since the test indicates that the KLIC with regard to model class $\mathcal{M}_1$ is significantly smaller than the KLIC of model class $\mathcal{M}_\in$ (compare to (9.5)).
- Similarly, we choose model class $\mathcal{M}_2$ over $\mathcal{M}_1$ if $\nu(\boldsymbol{x}_1, \ldots, \boldsymbol{x}_n) < -\Phi^{-1}\left(1 - \tfrac{\alpha}{2}\right)$.

Remark 9.9 (*Vuong tests for comparing vine copula models*) Up to now we introduced the general concept of the Vuong test. We now consider as model classes two competing regular vine copula classes. In general, they are non-nested model classes and thus Theorem 9.7 is applicable. This is the case, when they do not share pair copula terms and will be the case if the tree T_1 of the two vine tree structures does not share any edge. The case where they share pair copula terms is not easy to treat, but some general results for this overlapping case are available (see Theorem 6.3 of Vuong 1989).

For the case of nested models the standard likelihood ratio test is available as pointed out by Vuong (1989) in Sect. 7 of the paper. For regular vine copula models we have nested models, when one compares a regular vine copula model with one, where the same vine tree structure, the same pair copula families and the same parameters is used for the first m trees, while the remaining trees T_{m+1} to T_{d-1} are specified using the independence copula for each pair copula term. We speak of *truncated vine models*, which have been studied in Brechmann et al. (2012). Since the asymptotic distribution of the standard likelihood ratio test is not tractable, they apply the same Vuong test as suggested for the non-nested case. They justify their approach with a large simulation study, that proceeding this way gives satisfactory performance in finite samples.

9.3.1 Correction Factors in the Vuong Test for Adjusting for Model Complexity

The test defined in (9.8) does however not take into account the possibly different number of parameters of both models. Thus the model complexity as measured by the number of parameters is ignored. Hence the test is called *unadjusted* and Vuong (1989) gives the definition of an adjusted statistic

$$\widetilde{LR}_n(\hat{\theta}_1, \hat{\theta}_2)(X_1, \ldots, X_n) := LR_n(\hat{\theta}_1, \hat{\theta}_2)(X_1, \ldots, X_n) - K_n(f_1, f_2),$$

where $K_n(f_1, f_2)$ is the correction factor, which depends on the characteristics of the competing model classes $\mathcal{M}_1$ and $\mathcal{M}_2$ and is assumed to satisfy

$$n^{-1/2} K_n(f_1, f_2) = o_p(1). \tag{9.9}$$

Recall that $Z_n = o_p(1)$ for a set of random variables $(Z_n)_{n \in \mathbb{N}}$ corresponds to convergence to zero in probability, i.e., $\lim_{n \to \infty} P(|Z_n| > \varepsilon) = 0 \; \forall \varepsilon > 0$. Then the asymptotic result (9.7) also holds for $\widetilde{LR}_n$ and we can redefine test (9.8) in terms of $\widetilde{LR}_n$. In particular, Vuong (1989) notes that in terms of the unadjusted LR statistics, we would fail to reject the null hypothesis specified in (9.5) at level α if and only if

$$-\Phi^{-1}\left(1 - \frac{\alpha}{2}\right) + \frac{K_n(f_1, f_2)}{\sqrt{n\hat{\omega}^2}} \leq \nu(x_1, \ldots, x_n) \leq \Phi^{-1}\left(1 - \frac{\alpha}{2}\right) + \frac{K_n(f_1, f_2)}{\sqrt{n\hat{\omega}^2}}. \tag{9.10}$$

Here, ν is the scaled likelihood ratio statistics defined in (9.7).

Corresponding to the definition of AIC in (9.2) and BIC in (9.3), respectively, we consider the following two model complexity corrections satisfying (9.9) suggested by Vuong (1989):

- *Akaike correction*: $K_n^A(f_1, f_2) = k_1 - k_2$,
- *Schwarz correction*: $K_n^S(f_1, f_2) = \left(\frac{k_1}{2}\right)\ln(n) - \left(\frac{k_2}{2}\right)\ln(n)$,

where k_1 and k_2 denote the number of parameters in model class $\mathcal{M}_1$ and $\mathcal{M}_2$, respectively. The Schwarz correction again leads to more parsimonious models. Note however, that these choices were made out of convenience and other choices might be more appropriate depending on the specific setting, since there is a wide range of possible correction factors.

Remark 9.10 (Limits for the use of the Vuong test, AIC and BIC) Moreover, the Vuong test, as well as the AIC and BIC, only allows for a model selection among a set of possible models. If this model set is chosen badly, the selected "best" model can still be far off from the true model. The set of possible models therefore has to be carefully chosen.

Ex 9.3 (Model comparison of selected vine copula models for the extended WINE data set) We continue now our analysis of the data set considered in Exercise 1.7. Table 9.3 gives the results for all pairwise comparisons of the considered four models without utilizing the independence test (2.8). For this the function, RVineVuongTest of the VineCopula library by Schepsmeier et al. (2018) has been applied. We used joint maximum likelihood estimates of all copula parameters. Since all selected vine copula models have the same number of parameters, as can be seen from Table 9.2, no correction is used. We see that the R-vine copula model is better than the D-vine and Gaussian copula models, while the difference between the C-vine copula and R-vine copula is not significant at the 5% level. Also we have significant evidence to distinguish between C- and D-vines copula models.

Finally, we look at the comparison of the full specification and the one obtained from applying the pairwise test (2.8) for zero Kendall's τ values. Here, the reduced models have less parameters than the full specification, thus we consider also the Akaike and Schwarz adjustment. The results are contained in Table 9.4. From this, we see that there is no significant evidence to distinguish between the full and the reduced specification for the fitted regular vine copulas, Gaussian vine copulas and D-vine copulas, when the unadjusted or the Akaike adjusted Vuong test is performed. Thus, we can use the reduced model specification. For the C-vine copula specification, we have to stick with the full specification. The Schwarz correction is a larger correction than the Akaike correction, thus the acceptance region specified in (9.10) is shifted to right more than the one of the Akaike. Hence the adjusted LR statistics will become smaller, thus the p-values will become smaller. Thus under the Schwarz adjustment only the reduced R-vine copula and the D-vine copula can be utilized, while for all other vine copula models one has to stick with the full specification.

9.4 Exercises

Exer 9.1

ABALONE6: Comparing vine copula models for the six dimensional abalone *data set*: As in Exercise 8.1, we consider the variables length, diameter,

Table 9.3 WINE7: Comparing different fitted vine copula models using the asymptotic Vuong test in (9.8) with no adjustment

Comparison	Unadjusted LR statistics	p-value
R-vine versus C-vine	2.00	0.07
R-vine versus D-vine	4.00	0.00
R-vine versus Gauss-vine	7.83	0.00
C-vine versus D-vine	2.10	0.04
C-vine versus Gauss-vine	6.41	0.00
D-vine versus Gauss-vine	6.41	0.00

Table 9.4 WINE7: Comparing full and reduced by independence tests vine copula models (models ending with ind) using the Vuong test with Akaike and Schwarz corrections

Comparison	Unadjusted		Adjusted Akaike		Adjusted Schwarz	
	Statistic	p-value	Statistic	p-value	Statistic	p-value
R-vine versus R-vine-ind	1.61	0.11	0.82	0.41	−1.29	0.20
Gauss-vine versus Gauss-vine-ind	1.49	0.14	0.53	0.60	−2.06	0.04
C-vine versus C-vine-ind	3.07	0.00	2.92	0.00	2.53	0.00
D-vine versus D-vine-ind	0.87	0.38	0.31	0.76	−1.20	0.23

whole, shucked viscera and shell of the abalone data set contained in the R package PivotalR from Pivotal Inc. (2017). Again we restrict to female abalone shells and transform the data to the copula scale by using empirical distribution functions. Fit the following selected vine copula models using the Dißmann Algorithm 8.2.

- An R-vine, C-vine and D-vine copula using sequential and joint estimation of all copula parameter without applying the zero Kendall's τ test (2.8) allowing for all implemented pair copula families in the R-package VineCopula of Schepsmeier et al. (2018).
- An R-vine, C-vine and D-vine with applying a zero Kendall's τ test given in (2.8) and all other specifications as above.

For these fitted vine copula models perform the following model comparisons:

- Using the Akaike and BIC information criteria specified in (9.2) and (9.3), respectively, find the best fitting R-vine copula model among all above fitted R-vine models. Do the same for the fitted C- and D-vine copula models. Finally, compare between the different vine copula classes.
- Using the likelihood ratio test (9.8), compare first the models within their vine copula class and then compare between the best fitting member of the vine copula class.
- Given the comparison results, which model would you choose?

Exer 9.2

ABALONE6: *Finding more parsimonious vine copula models for the six dimensional* abalone *data set*: Again, consider the data set of Exercise 9.1. We now want to investigate, if one can reduce the choice of pair copula families without loosing goodness of fit. For this perform, the same tasks as in Exercise 9.1 for each specified family set below:

- Only allowing a single parametric bivariate copula family. Do this separately for the Gaussian, Student's t and Gumbel copula family.
- Only allowing the implemented parametric bivariate copula families with a single parameter.
- The bivariate Gaussian, Student's t or the Gumbel copula (including rotations)
- Only bivariate Gaussian copulas resulting in a Gaussian vine.

Given your results, which vine copula specification would you choose?

Exer 9.3

URANIUM7: *Comparing vine copula models for the seven dimensional* uranium *data set*: Consider now all variables of the uranium data set contained in the R package copula of Hofert et al. (2017). Perform a similar analysis as in Exercise 9.1 and 9.2.

Case Study: Dependence Among German DAX Stocks

10

10.1 Data Description and Sector Groupings

Understanding dependence among financial stocks is vital for option pricing and forecasting portfolio returns. Copula modeling has a long history in this area. However, the restrictive use of the multivariate Gaussian copula with no tail dependence has been blamed for the financial crisis in 2007–2008 (see Li 2000, Salmon 2012 and recently Puccetti and Scherer 2018), and therefore it is important to allow for much more flexible dependence models such as allowed by the vine copula class. In this context, the possibility of modeling tail dependence by vine copulas has to be mentioned.

Since financial data are strongly dependent on past values, copula-based models have to account for this. Therefore, a two-step inference for margins approach is appropriate. This allows us to remove the marginal time dependence by utilizing standard time series models and to consider the dependence structure among standardized residuals. In particular, generalized autoregressive conditional heteroskedasticity (GARCH) models are commonly used in this context. They allow for time-varying volatility and volatility clustering, which are often observed in financial data. Roughly speaking, they model the variance at time t as function of the past by including the variance and the observed value at time $t - 1$. Thus, they are observation-driven dynamic models, which allow for forecasting. They are described in much more detail in the book of Francq and Zakoian (2011).

In this case study, we investigate the dependence structure among the daily values of German stocks included in the German financial index DAX from January 4, 2005 until July 22, 2011 after we have removed the marginal time dependence. The DAX (Deutscher Aktienindex (German stock index)) is a blue chip stock market index consisting of the 30 major German companies trading on the Frankfurt Stock Exchange. Prices are taken from the Xetra trading venue. According to Deutsche Börse, the operator of Xetra, DAX measures the performance of the Prime

© Springer Nature Switzerland AG 2019
C. Czado, *Analyzing Dependent Data with Vine Copulas*, Lecture Notes in Statistics 222, https://doi.org/10.1007/978-3-030-13785-4_10

Standard's 30 largest German companies in terms of order book volume and market capitalization. It is the equivalent of the FT 30 and the Dow Jones Industrial Average.

There are 30 stocks in the DAX to consider. To keep this case study manageable, we do not study the dependence structure among all stocks of the DAX. We restrict to the analysis of the dependence among sectors. For this, we select a representative for each sector and build up a vine copula to model the dependence. The corresponding ticker names, the companies, and their sector are given in Table 10.1.

Table 10.1 DAX: Ticker and company name and the corresponding sector

Ticker name	Company	Sector
ADS.DE	ADIDAS	retail
ALV.DE	ALLIANZ	financial
BAS.DE	BASF	chem.health
BAYN.DE	BAYER	chem.health
BEI.DE	BEIERSDORF	retail
BMW.DE	BMW	automobile
CBK.DE	COMMERZBANK	financial
DAI.DE	DAIMLER	automobile
DB1.DE	DEUTSCHE.BOERSE	financial
DBK.DE	DEUTSCHE.BANK	financial
DPW.DE	DEUTSCHE.POST	trans.util.it
DTE.DE	DEUTSCHE.TELEKOM	trans.util.it
EOAN.DE	E.ON	trans.util.it
FME.DE	FRESENIUS.Med.Care	chem.health
FRE.DE	FRESENIUS	chem.health
HEI.DE	HEIDELBERGCEMENT	industrial
HEN3.DE	HENKEL	retail
IFX.DE	INFINEON	trans.util.it
LHA.DE	DT.LUFTHANSA	trans.util.it
LIN.DE	LINDE	chem.health
MAN.DE	MAN	industrial
MEO.DE	METRO	retail
MRK.DE	MERCK	chem.health
MUV2.DE	MUNICHRE	financial
RWE.DE	RWE	trans.util.it
SAP.DE	SAP	trans.util.it
SDF.DE	K.S	chem.health
SIE.DE	SIEMENS	industrial
TKA.DE	THYSSENKRUPP	industrial
VOW3.DE	VOLKSWAGEN	automobile

10.2 Marginal Models

Often the observed sample does not constitute an independent identically distributed sample. This is however needed, if we want to use the estimation techniques discussed in Chap. 7. In particular, for time series data as being considered here, we need as already mentioned to remove the serial dependence present in each component. This will be accomplished by using standard univariate financial time series models such as the class of GARCH models.

For each of the 30 stocks, we fit a $GARCH(1, 1)$ model with standardized Student t innovations using the function $\texttt{garchFit}$ of the R library $\texttt{fgarch}$ (Wuertz et al. 2017). A $GARCH(1, 1)$ model for a time series $Y_t, t = 1, \ldots, T$ is defined using the conditional variance $\sigma_t := Var(Y_t|Y_1, \ldots, Y_{t-1})$ and innovation variables $Z_t, t = 1, \ldots, T$ by

$$Y_t = \sigma_t Z_t$$
$$\sigma_t = \omega + \alpha Y_{t-1}^2 + \beta \sigma_{t-1}^2. \tag{10.1}$$

For our data, we assumed the distribution of Z_t as standardized Student t. A standardized Student t distribution is a scaled Student t distribution with zero mean and unit variance. As a check that the resulting standardized innovations estimates are a random sample from the standardized Student t distribution, we use a two-sided Kolmogorov–Smirnov test (Massey Jr 1951). The resulting p-values are given in Fig. 10.1. This shows that nearly for all stocks the assumption of a standardized Student t-distribution cannot be rejected. Therefore, we use the cumulative distribution function of the standardized Student t distribution to define the copula data as probability integral transformation, i.e., we determine

$$u_{it} := F(\frac{y_{it}}{\hat{\sigma}_{it}}; \hat{\nu}_i),$$

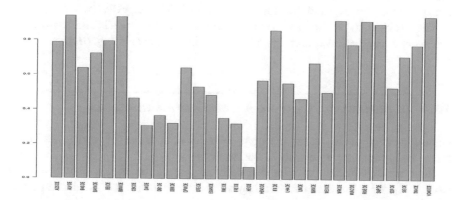

Fig. 10.1 DAX: P-values of Kolmogorov–Smirnov tests for independence of each DAX stock

where $F(\cdot; \hat{\nu}_i)$ is the distribution function of the innovation distribution with estimated degree of freedom $\hat{\nu}_i$ and estimated conditional variance $\hat{\sigma}_{it}^2$ for stock $i = 1, \ldots, 30$ at time $t = 1, \ldots, T$.

10.3 Finding Representatives of Sectors

First, we illustrate the dependencies within sectors through pairs plots and normalized contour plots. The corresponding plots for each of the six sectors are contained in Fig. 10.2. For the automobile sector, we see that DAIMLER (DAI.DE) has strong dependence with the other two automobile stocks. The pairwise empirical Kendall's τ values range from .34 to .51. For the chem/health sector, we also see positive dependence (Kendall's τ range from .13 until .43) with stronger dependencies to BASF (BAS.DE). For the financial sector, we identify strong positive dependencies with Kendall's τ estimates between .31 and .52. In the industrial sector, dependencies are less elliptical with Kendall's τ from .27 to .51, while the transport/utility/energy sector has stronger elliptical dependencies with Kendall's τ from .21 until .58. The dependencies among the retail stocks are also elliptical with pairwise Kendall's τ between .25 and .29.

C-vine tree structures are ideal, when we are interested in finding a ranking of the nodes according to the strength of the dependence as measured by the Kendall's τ. We use this property to find representative companies for each sector. Table 10.2 gives the number of companies grouped in one sector as well as the root node of a C-Vine fit within each sector. These root node companies we choose as representative of the corresponding sector. In particular, DAIMLER (DAI.DE) is selected as proxy of the automotive industry in the DAX, while BASF (BAS.DE) represents the chemical/health industry. The financial industry is represented by the DEUTSCHE BANK (DKB.DE), while SIEMENS (SIE.DE) is chosen as proxy of the industrial sector. RWE (RWE.DE) and HENKEL (HEN3.DE) are the representatives of the transport/utility/energy and the retail sector in the DAX, respectively.

The resulting first tree of the C-vine for each sector chem.health, financial, industrial, retail, and trans.util.it are given in Fig. 10.3.

10.4 Dependence Structure Among Representatives

Having selected the representative companies for each of the six sectors, we investigate the dependence among those companies.

First, we explore the pairwise dependencies among the representatives of each stock in Fig. 10.4. From the contour shapes, we see evidence of positive dependencies and the shapes suggest the presence of the Student t copula for bivariate pairs. The Kendall's τ estimates a range between .22 and .46. To model these dependencies, we allow for R-, C-, and D-vine copulas. The function RVineStructureSelect

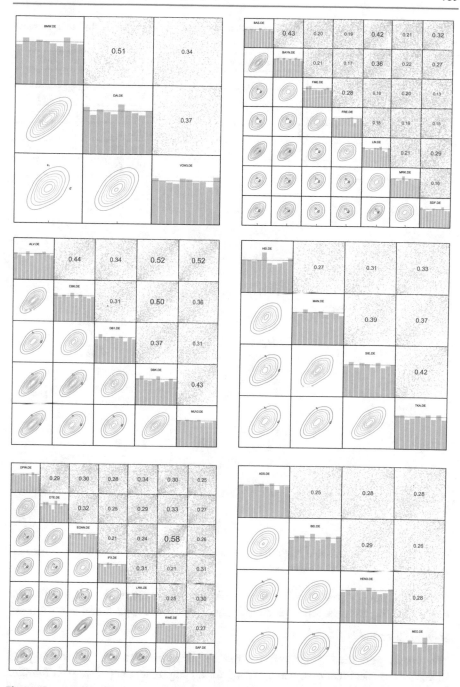

Fig. 10.2 DAX: Pairs plots for automotive, chemical/health (top row), financial, industrial (middle row), transport/ultility/energy, and retail sector (bottom row) (Within each panel: Upper: pairs plots of copula data, diagonal: histogram of copula margins, lower: normalized contour plots)

Table 10.2 DAX: Number of companies grouped to the six sectors together with their representative company

Sector	Number of companies	Representative company
automobile	3	DAI.DE
chem.health	7	BAS.DE
financial	5	DBK.DE
industrial	4	SIE.DE
trans.util.it	7	RWE.DE
retail	4	HEN3.DE

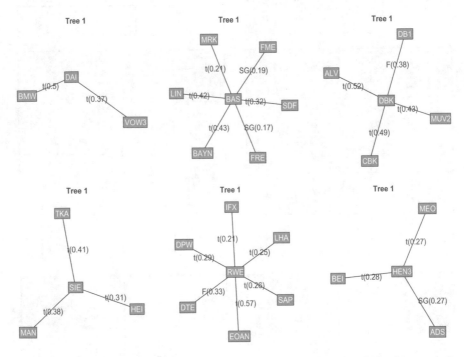

Fig. 10.3 DAX: First C-vine trees for the sectors `auto`, `chem.health`, `financial`, `industrial`, `trans.util.it` and `retail`

of the R library `VineCopula` of Schepsmeier et al. (2018) can be applied for accomplishing this task. All implemented parametric pair copula families are utilized.

R-vine Copulas

We present now the regular vine fits for the six representatives. We investigated three R-vine specifications, one allowing for all implemented pair copula families (`rv`), one where we only allowed Gaussian and Student t copulas (`rvt`), and one where only Gaussian pair copulas are allowed (`rvg`). Note that the last specification corresponds to a multivariate Gaussian copula with correlations specified by the corresponding partial correlations.

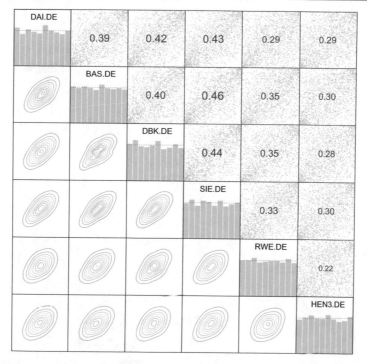

Fig. 10.4 DAX: Upper: pairs plots of copula data, diagonal: histogram of copula margins, lower: normalized contour plots for the representative companies of the sectors

Additionally, we allow for an asymptotic $\alpha = .05$ level independence test based on Kendall's τ introduced in (2.8). If this test is nonsignificant for a pair copula term, then we set the associated family to the independence copula. This allows us to reduce the complexity of the model.

Parameters are estimated using the sequential estimation method and the joint maximum likelihood discussed in Sect. 7.2. The estimates in Fig. 10.5 are based on sequential estimates.

The first two rows of Fig. 10.5 give the selected R-vine tree structure together with the selected pair copula family and the fitted Kendall's τ value, when all pair copula families are allowed. We see that in the tree T_1, all pair copula families are selected as Student t copulas. This has been often observed in financial data sets. The trees T_j with $j > 1$ however also include non-elliptical pair copula families, but the strength as measured by the fitted Kendall's τ values is quite weak.

The last two rows of Fig. 10.5 give the corresponding results, when only Student t and Gaussian pair copulas are allowed. We observe no change in the selected vine tree structure, but changes to the selection of the pair copula families and thus to the associated parameter estimates in trees T_j for $j \geq 2$. Since here the fitted Kendall's τ values are quite low, the difference between the fitted models will not be large.

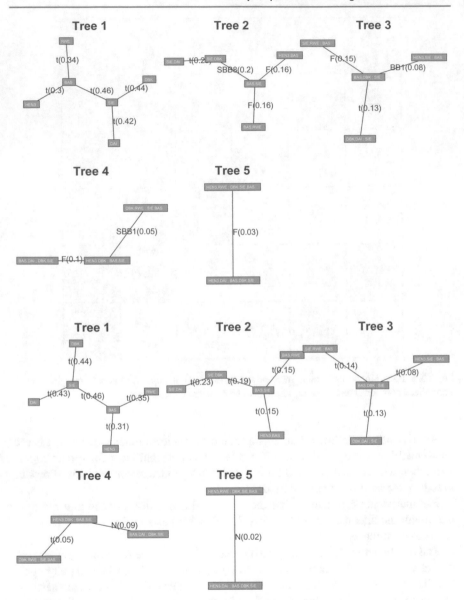

Fig. 10.5 DAX: R-vine trees allowing for all pair copula families (top) and allowing only for Student t or Gaussian copulas (bottom) for sector representatives together with their pair copula family and the fitted Kendall's τ value for each edge based on joint maximum likelihood

To investigate more precisely the influence of the selected pair copula family, we study the associated fitted normalized contour plots given in Fig. 10.6. Here, the selection allowed all implemented pair copula families. From this, we clearly see the fitted diamond-shaped contours associated with the bivariate Student t copulas.

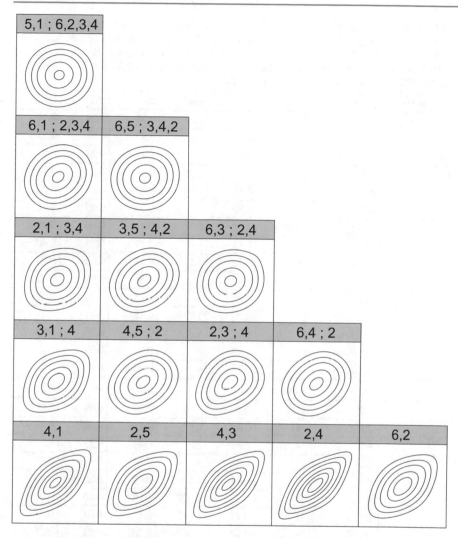

Fig. 10.6 DAX: Fitted R-vine normalized contour for sector representatives allowing for all pair copulas (rv) based on joint maximum likelihood (node abbreviations: 1 <-> DAI.DE, 2 <-> BAS.DE, 3 <-> DBK.DE, 4 <-> SIE.DE, 5 <-> RWE.DE, 6 <-> HEN3.DE)

Some nonsymmetry is visible in contour shapes in trees T_j for $j \geq 2$. One example for this behavior is the fitted conditional pair copula $C_{23;4}$. This property cannot be modeled by bivariate Student t copulas.

For this data set, we noticed that the asymptotic α level test for independence given in (2.8) does not give more parsimonious R-vine specifications, thus we do not present detailed summaries for these fits. Further, we do not show the detailed results for the joint maximum likelihood estimation of the R-vine copula since they are very close to the sequential estimates.

Table 10.3 DAX: Sequential parameter estimates (par and if necessary par2) together with implied Kendall's τ (tau), upper (utd) and lower (ltd) tail dependence coefficient estimates for the R-vine copula (allowing for all implemented pair copula families), and the R-vine copula with only Student t and Gaussian pair copulas (node abbreviations: 1 <-> DAI.DE, 2 <-> BAS.DE, 3 <-> DBK.DE, 4 <-> SIE.DE, 5 <-> RWE.DE, 6 <-> HEN3.DE)

R-vine tree	edge	cop	par	par2	tau	utd	ltd
1	6,2	t	0.45	8.32	0.30	0.09	0.09
1	2,4	t	0.66	4.47	0.46	0.33	0.33
1	4,3	t	0.63	5.56	0.44	0.27	0.27
1	2,5	t	0.51	6.37	0.34	0.17	0.17
1	4,1	t	0.61	5.68	0.42	0.25	0.25
2	6,4;2	F	1.50	–	0.16	0.00	0.00
2	2,3;4	SBB8	2.33	0.71	0.20	0.00	0.00
2	4,5;2	F	1.53	–	0.17	0.00	0.00
2	3,1;4	t	0.35	9.19	0.23	0.05	0.05
3	6,3;2,4	BB1	0.07	1.05	0.08	0.06	0.00
3	3,5;4,2	F	1.36	–	0.15	0.00	0.00
3	2,1;3,4	t	0.20	12.98	0.13	0.01	0.01
4	6,5;3,4,2	SBB1	0.07	1.01	0.05	0.00	0.02
4	6,1;2,3,4	F	0.93	–	0.10	0.00	0.00
5	5,1;6,2,3,4	F	0.25	–	0.03	0.00	0.00
t R-vine tree	edge	cop	par	par2	tau	utd	ltd
1	6,2	t	0.45	8.32	0.30	0.09	0.09
1	2,4	t	0.66	4.47	0.46	0.33	0.33
1	4,3	t	0.63	5.56	0.44	0.27	0.27
1	2,5	t	0.51	6.37	0.34	0.17	0.17
1	4,1	t	0.61	5.68	0.42	0.25	0.25
2	6,4;2	t	0.23	21.27	0.15	0.00	0.00
2	2,3;4	t	0.29	14.42	0.19	0.01	0.01
2	4,5;2	t	0.23	12.07	0.15	0.01	0.01
2	3,1;4	t	0.35	9.19	0.23	0.05	0.05
3	6,3;2,4	t	0.13	17.72	0.08	0.00	0.00
3	3,5;4,2	t	0.22	14.83	0.14	0.01	0.01
3	2,1;3,4	t	0.20	13.77	0.13	0.01	0.01
4	6,5;3,4,2	t	0.08	30.00	0.05	0.00	0.00
4	6,1;2,3,4	N	0.14	-	0.09	0.00	0.00
5	5,1;6,2,3,4	N	0.04	–	0.02	0.00	0.00

In the top part of Table 10.3, we present the sequential parameter estimates of the R-vine specification allowing for all implemented pair copula families. From this, we see that the bivariate Student t copula is most often selected and especially it is selected for all edges in the tree T_1. The other selected pair copula families imply no or only little tail dependence. The first tree T_1 identifies Siemens (SIE.DE) from the industrial sector and BASF (BAS.DE) from the chemical/heath sector as central nodes in the dependence network identified by the R-vine structure.

The corresponding results for the R-vine, where only bivariate Student t copulas are allowed for the pair copula families, are contained in the bottom part of Table 10.3. The estimated degree of freedom parameters in the first tree varies between 4.47 and 8.32, thus indicating different tail behaviors of pairs, which cannot be captured by a six-dimensional Student t copula, which allows only for a single degree of freedom parameter. In trees T_j for $j \geq 2$ larger estimated degree of freedom parameters are observed, thus a decrease in tail dependence is fitted for these conditional pairs. For two pair copula terms in trees T_4 and T_5 even the Student t copulas is replaced by the Gaussian copula.

C- and D-vine Copulas

Similarly, we conduct a dependence analysis using C- and D-vine tree sequences. We denote these specifications by cv and dv, respectively. Again the selected tree structure, the selected pair copula families, and the induced Kendall's τ values by sequential estimation are displayed in Fig. 10.7.

For the C-vine specification, the root node ordering is Siemens (SIE.DE) from the industrial sector, Deutsche Bank (DBK.DE) from the financial sector, BASF (BAS.DE) from the chemical/health sector, Henkel (HEN3.DE) from the retail sector, RWE (RWE.DE) from the transport/utility/IT sector, and Daimler (DAI.DE) from the automotive sector. This gives an estimated importance ordering of the sector representatives.

The D-vine specification is most appropriate, if there is a prior ordering of the variables available. This is not the case for this data set. The fitted D-vine assigns the ordering RWE- Henkel- Daimler- Deutsche Bank- Siemens- BASF to the sector representatives. Looking at the fitted Kendall's τ, we see that they are increasing with the ordering. This allows for the following interpretation: BASF is most dependent on Siemens, Siemens on Deutsche Bank after the effect of BASF is accounted for, Deutsche Bank on Daimler after BASF and Siemens are accounted, Daimler on Henkel after BASF, Siemens and Deutsche Bank are accounted and Henkel on RWE after BASF Siemens, Deutsche Bank and Daimler are accounted.

Again, we explore the normalized contour shapes associated with the C- (top) and D-vine (bottom) fits in Fig. 10.8. A nonsymmetric pair copula is already fitted in tree T_1 for the C-vine specification.

Now, we present the summary tables of the fitted C- and D-vine copulas using the sequential estimation method and allowing for all implemented pair copula families. They are contained in Table 10.4. We see that the selected C-vine copula has already a non-elliptical pair copula family in tree T_1, while this is not the case for the chosen D-vine copula. In general, there are more non-elliptical pair copulas chosen for these more restricted R-vine copula specifications. Therefore, we will compare all studied model classes in the next section using the methods discussed in Chap. 9.

Table 10.4 DAX: Sequential parameter estimates (par and if necessary par2) together with implied Kendall's τ (tau), upper (utd) and lower (ltd) tail dependence coefficients for the C- and D-vine specifications allowing all implemented pair copulas. (node abbreviations: 1 <-> DAI.DE, 2 <-> BAS.DE, 3 <-> DBK.DE, 4 <-> SIE.DE, 5 <-> RWE.DE, 6 <-> HEN3.DE)

C-vine tree	edge	cop	par	par2	tau	utd	ltd
1	6,4	SBB1	0.07	1.36	0.29	0.00	0.34
1	4,3	t	0.63	5.56	0.44	0.27	0.27
1	4,2	t	0.66	4.47	0.46	0.33	0.33
1	4,5	t	0.49	5.20	0.32	0.19	0.19
1	4,1	t	0.61	5.68	0.42	0.25	0.25
2	6,3;4	BB1	0.11	1.07	0.11	0.09	0.00
2	3,2;4	SBB8	2.33	0.71	0.20	0.00	0.00
2	3,5;4	F	1.91	–	0.21	0.00	0.00
2	3,1;4	t	0.35	9.19	0.23	0.05	0.05
3	6,2;3,4	F	1.25	–	0.14	0.00	0.00
3	2,5;3,4	F	1.49	–	0.16	0.00	0.00
3	2,1;3,4	t	0.20	12.98	0.13	0.01	0.01
4	6,5;2,3,4	G	1.05	–	0.04	0.06	0.00
4	6,1;2,3,4	F	0.99	–	0.11	0.00	0.00
5	5,1;6,2,3,4	F	0.30	–	0.03	0.00	0.00
D-vine tree	edge	cop	par	par2	tau	utd	ltd
1	2,4	t	0.66	4.47	0.46	0.33	0.33
1	4,3	t	0.63	5.56	0.44	0.27	0.27
1	3,1	t	0.61	6.03	0.42	0.23	0.23
1	1,6	t	0.44	11.77	0.29	0.04	0.04
1	6,5	t	0.34	9.26	0.22	0.05	0.05
2	2,3;4	SBB8	2.33	0.71	0.20	0.00	0.00
2	4,1;3	t	0.37	9.30	0.24	0.05	0.05
2	3,6;1	t	0.21	11.90	0.14	0.01	0.01
2	1,5;6	SBB8	3.12	0.60	0.23	0.00	0.00
3	2,1;4,3	t	0.20	13.15	0.13	0.01	0.01
3	4,6;3,1	F	1.29	–	0.14	0.00	0.00
3	3,5;1,6	F	2.03	–	0.22	0.00	0.00
4	2,6;4,3,1	F	1.04	–	0.11	0.00	0.00
4	4,5;3,1,6	SBB8	2.05	0.61	0.12	0.00	0.00
5	2,5;4,3,1,6	F	1.29	–	0.14	0.00	0.00

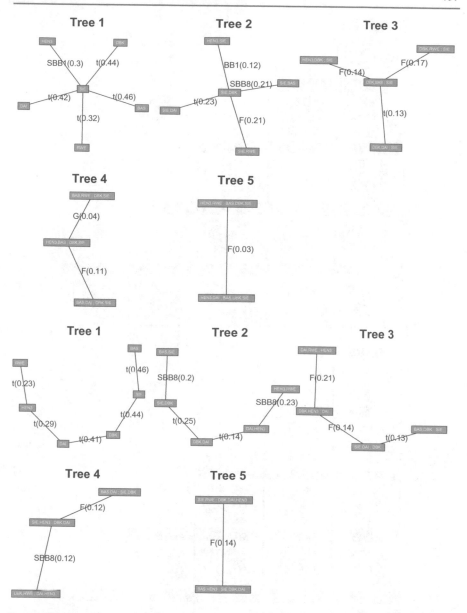

Fig. 10.7 DAX: C-vine (top) and D-vine (bottom) trees for sector representatives together with their chosen pair copula families and the corresponding fitted Kendall's τ values based on joint maximum likelihood estimation

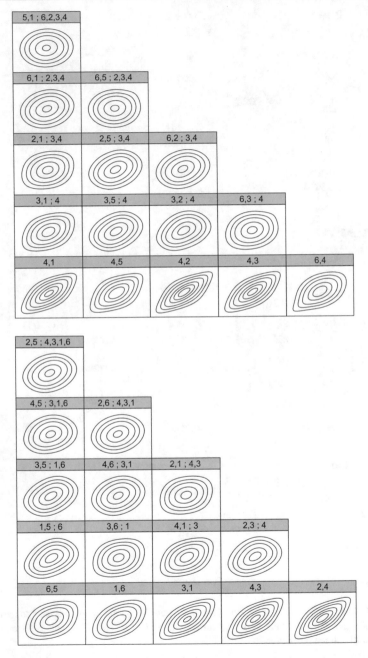

Fig. 10.8 DAX: Fitted C-vine (top) and D-vine (bottom) normalized contours for sector representatives allowing for all pair copulas based on joint maximum likelihood (node abbreviations: 1 <-> DAI.DE, 2 <-> BAS.DE, 3 <-> DBK.DE, 4 <-> SIE.DE, 5 <-> RWE.DE, 6 <-> HEN3.DE)

10.5 Model Comparison

We now compare the different models by studying the standard information-based model AIC and BIC criteria as well as the Vuong test discussed in Sect. 9.3 suitable for non-nested model comparison.

Information-Based Model Comparison

In Table 10.5, we give the estimated log-likelihood, the number of parameters, and the AIC and BIC statistics of all copula models studied. From this, we see that the sequential estimates nearly produce the same log-likelihood as the joint maximum

Table 10.5 DAX: Estimated log-likelihood (loglik), number of parameters (df), AIC and BIC for all fitted models (seq = sequential estimation, mle = maximum likelihood, ind = asymptotic α level independence tests used, all = all implemented pair copula families used, t = only Student t and Gaussian pair copulas used, G = only Gaussian copulas used)

G-vine-seq	loglik	par	AIC	BIC
	1952.74	15	−3875.48	−3794.28
R-vine-seq-all	2142.09	25	−4234.17	−4098.84
R-vine-mle-all	2144.25	25	−4238.50	−4103.16
R-vine-ind-seq-all	2140.81	24	−4233.62	−4103.70
R-vine-ind-mle-all	2142.87	24	−4237.75	−4107.83
R-vine-seq-t	2137.66	28	−4219.32	−4067.74
R-vine-mle-t	2140.49	28	−4224.99	−4073.41
R-vine-ind-seq-t	2136.54	27	−4219.08	−4072.92
R-vine-ind-mle-t	2139.30	27	−4224.59	−4078.43
C-vine-seq-all	2156.96	24	−4265.92	−4136.00
C-vine-mle-all	2160.71	24	−4273.43	−4143.51
C-vine-ind-seq-all	2156.96	24	−4265.92	−4136.00
C-vine-ind-mle-all	2160.71	24	−4273.43	−4143.51
C-vine-seq-t	2135.58	27	−4217.16	−4071.00
C-vine-mle-t	2139.03	27	−4224.05	−4077.89
C-vine-ind-seq-t	2134.44	26	−4216.89	−4076.14
C-vine-ind-mle-t	2137.81	26	−4223.61	−4082.86
D-vine-seq-all	2155.92	26	−4259.84	−4119.09
D-vine-mle-all	2159.25	26	−4266.49	−4125.74
D-vine-ind-seq-all	2155.92	26	−4259.84	−4119.09
D-vine-ind-mle-all	2159.25	26	−4266.49	−4125.74
D-vine-seq-t	2125.26	29	−4192.51	−4035.53
D-vine-mle-t	2128.77	29	−4199.54	−4042.55
D-vine-ind-seq-t	2125.26	29	−4192.51	−4035.53
D-vine-ind-mle-t	2128.77	29	−4199.54	−4042.55

likelihood estimates, while there is little reduction in the number of parameters when the asymptotic α level independence test (2.8) is invoked. Clearly, we see that the multivariate Gaussian copula is not sufficient, while AIC and BIC slightly prefer the C-vine model over the D-vine and R-vine model. We note that the estimation results do not change for the Gaussian specification, when joint maximum likelihood and/or the asymptotic α level independence test is applied. Therefore, we do not show them in Table 10.5.

Non-nested Model Comparison Using Vuong Tests

First, we compare the fitted models within each vine class. In particular, we want to investigate, if there is significant evidence for difference between the model specifications with regard to the pair copula families allowed and whether the asymptotic α level independence test (2.8) is applied. For this, we use the Vuong test discussed in Sect. 9.3 applied to the joint maximum likelihood estimates of the vine specification. The results are presented in Table 10.6. We see that all R-vine specifications cannot be distinguished by the Vuong test on the 5% level. For C-vines, the restriction to only Student t and Gauss copula families produces a significantly worse fit on the 5% level compared to the one, where all implemented pair copula fits are allowed. The same conclusions as for the C-vine specifications can be drawn for the D-vine specifications. So in the next comparison, we restrict ourselves to comparisons allowing for all implemented pair copula families and no asymptotic α level independence test.

To distinguish between the vine classes, we conduct again pairwise comparison using the Vuong test. The results are reported in Table 10.7. They show that there

Table 10.6 DAX: Vuong test model comparison within each vine class using joint maximum likelihood estimation

Comparison	Unadjusted		Akaike adjusted		Schwarz adjusted	
	Stat	p-value	Stat	p-value	Stat	p-value
rv all versus t	0.39	0.69	0.71	0.48	1.56	0.12
rv ind-all versus ind-t	0.38	0.70	0.70	0.49	1.56	0.12
rv all versus ind	0.80	0.42	0.22	0.83	−1.36	0.17
rv t versus ind-t	0.78	0.44	0.13	0.90	−1.63	0.10
cv all versus t	2.13	0.03	2.43	0.02	3.23	0.00
cv ind-all versus ind-t	2.22	0.03	2.41	0.02	2.94	0.00
cv all versus ind	0.00	1.00	0.00	1.00	0.00	1.00
cv t versus ind t	0.79	0.43	0.14	0.89	−1.60	0.11
dv all versus t	2.59	0.01	2.84	0.00	3.53	0.00
dv ind-all versus ind-t	2.59	0.01	2.84	0.00	3.53	0.00
dv all versus ind	0.00	1.00	0.00	1.00	0.00	1.00
dv t versus ind-t	0.00	1.00	0.00	1.00	0.00	1.00

Table 10.7 DAX: Vuong test model comparison between each vine class using joint maximum likelihood estimation allowing for all implemented pair copula families and without the independence test (2.8)

Comparison	Unadjusted		Akaike adjusted		Schwarz adjusted	
	Stat	p-value	Stat	p-value	Stat	p-value
rv all versus dv all	−1.44	0.15	−1.53	0.13	−1.77	0.08
rv all versus cv all	−1.21	0.23	−1.13	0.26	−0.91	0.36
cv all versus dv all	0.13	0.89	0.31	0.75	0.80	0.42
rv all versus gv	6.68	0.00	6.33	0.00	5.39	0.00

is no significant evidence on the 5% level that the vine class specifications provide a different fit, while there is strong statistical evidence that the Gaussian copula provides an inferior fit compared to the vine copulas.

Finally, we like to note that this comparison is all in sample. A more rigorous comparison would involve cross-validation.

10.6 Some Interpretive Remarks

This analysis shows that the class of vine models can more appropriately explain the dependence structure among sector representatives of the DAX index than a Gaussian copula. The stock from SIEMENS (SIE.DE) is selected as the most important stock among the six sector representatives. It is followed by DEUTSCHE BANK (DKB.DE), BASF (BAS.DE), DAIMLER (DAI.DE), HENKEL(HENS3.DE), and RWE (RWE.DE). This indicates that the dependency among German economy is driven by industrial and financial companies.

Finding a good model for the dependence structure is only one building block for the financial risk analyst. For an introduction to the methods of risk management see the book by McNeil et al. (2015). Such a person is especially interested in forecasting the value at risk of a portfolio of interest. The value of risk is a high quantile of the portfolio return. For such an analysis, we have to simulate a large sample from the fitted vine copula model and transfer the sample to the standardized residual level using for each component the inverse of the fitted residual distribution. For this step, an invertible fitted marginal distribution is necessary. The univariate empirical distribution is not suitable for this task.

The sample of the standardized residuals is then used to construct a forecast sample of the portfolio return using the dynamic marginal GARCH models. Finally, the forecast sample of the portfolio return is then taken to construct estimates of the value at risk. This way of proceeding has been followed and is illustrated using vine copula models in Brechmann and Czado (2013) for the Euro Stoxx 50 index.

Recent Developments in Vine Copula Based Modeling

11

So far, we have given a basic introduction to vine copulas with some applications. The field is rapidly developing and in this chapter we showcase some developments in the area of estimation, model selection and adaptions to special data structures. Next, we summarize current applications to problems from finance, life and earth sciences. The chapter closes with a section on available software to conduct vine copula based modeling.

11.1 Advances in Estimation

Bayesian Vine Copula Analysis

Estimation of vine copula parameters so far was treated in a frequentist way. In a frequentist perspective unknown parameters are considered as fixed quantities and estimation are commonly performed by maximum likelihood. In contrast, the Bayesian framework allows unknown parameters to be random quantities and prior knowledge about them are summarized in a prior distribution. We denote the associated prior density for the random parameter vector θ by $p(\theta|D)$. After seeing the data the knowledge of the parameter is updated by determining the posterior distribution of parameters. This is defined as the distribution of the parameter vector θ given the data D. The corresponding density will be denoted by $p(\theta|D)$. The posterior density is then proportional to the likelihood $l(\theta; D)$ times the prior, i.e., $p(\theta|D) \propto l(\theta; D) \cdot \pi(\theta)$. Since only in special cases the posterior distribution is known analytically, it is often approximated by Markov Chain Monte Carlo (MCMC) methods. For more details, contact the books by Gilks et al. (1995) and Robert (2004).

The first analysis of vine copulas in a Bayesian framework was conducted by Min and Czado (2010). They analyze a D-vine copula, where the pair copulas of the vine are bivariate Student t copulas. Parameters of this model are estimated

© Springer Nature Switzerland AG 2019
C. Czado, *Analyzing Dependent Data with Vine Copulas*, Lecture Notes
in Statistics 222, https://doi.org/10.1007/978-3-030-13785-4_11

using MCMC by adapting the Metropolis–Hastings algorithm of Metropolis et al. (1953). The MCMC algorithm provides credible intervals for estimated parameters. In contrast, confidence intervals are difficult to obtain for vine copulas, if maximum likelihood estimation is used. Following to this, Min and Czado (2011) employ Bayesian model selection in D-vine copulas to decide, whether a pair copula is the independence copula or not. This is achieved by introducing model indicators and utilizing reversible jump MCMC approaches of Green (1995). This approach leads to a more parsimonious model compared to the full D-vine specification. Smith et al. (2010) also employ Bayesian model selection for a D-vine and use this method to model serial dependence within longitudinal data. Smith (2015) models both the serial and the cross-sectional dependence of a multivariate time series with a D-vine copula and provides an MCMC algorithm for parameter inference.

So far Bayesian model selection was restricted to a subclass of regular vine copulas. Gruber and Czado (2015) provide a sequential method based also on reversible jump MCMC, which is able to deal with regular vine copulas in general. This method selects sequentially the vine tree structure and corresponding pair copula families from a given set of copula families. In a simulation study, it is shown that this method leads to better results compared to the standard frequentist model selection Algorithm 8.2 of Dißmann et al. (2013). However, this is achieved at higher computational costs. In a follow-up paper Gruber and Czado (2018) developed a non-sequential Bayesian method for the vine structure by choosing the tree to be updated randomly. They showed that the associated MCMC sampler mixes well in small dimensions and give an application to financial data.

For factor vine copula models the Bayesian approach also seems to be promising and provides advantages as further discussed in Sect. 11.3. In fact, Schamberger et al. (2017) develop a first Bayesian approach for a one-factor vine copula model. They follow a two step approach to facilitate inference for the marginal and copula parameters. Very recently Kreuzer and Czado (2018) are able to develop a joint Bayesian inference approach using Hamiltonian Monte Carlo of Hoffman and Gelman (2014) for a one-factor stochastic volatility model.

So far we have discussed Bayesian approaches to infer vine copula parameters. Vine copulas can also be helpful to design or improve an inference procedure itself. Schmidl et al. (2013) show, how vine copulas can be used to obtain multivariate proposals for the Metropolis–Hastings algorithm. The vine copula based proposal incorporates the dependency between model parameters, which leads to an improved MCMC procedure. They apply this approach to get better mixing over standard approaches in inference problems in system biology.

Vine Copula Models with Nonparametric Pair Copulas

In Sect. 3.8, we introduced the empirical normalized contour plot as visual guide to check the appropriateness of a bivariate parametric copula family. But sometimes none of the bivariate copula families discussed in Chap. 3 seems to fit the data. As such a data example, Fig. 11.1 shows several pairs from the MAGIC telescope data set (https://archive.ics.uci.edu/ml/datasets/magic+gamma+telescope). The rank transformed observations (first row) indicate that the dependence has rather uncommon characteristics, which is confirmed by the empirical normalized contour plots (second row). The contour shapes do not correspond to any of the parametric copula families introduced in Chap. 3. For such situations, a nonparametric approach is preferable.

Several specialized methods for bivariate copula densities were proposed in the literature. One technique was already mentioned in Sect. 3.8. The idea is to transform the copula data to the z-scale by setting $(z_{i1}, z_{i2}) := (\Phi^{-1}(u_{i1}), \Phi^{-1}(u_{i2}))$ for $i = 1, \ldots, n$. We now estimate the density $g(z_1, z_2)$ of the associated random vector $(Z_1, Z_2) := (\Phi^{-1}(U_1), \Phi^{-1}(U_2))$ by a kernel estimator. An estimate of the copula density c is then obtained by rescaling the kernel density estimate $\widehat{g}$ to uniform margins by

$$\widehat{c}(u_1, u_2) := \frac{\widehat{g}\big(\Phi^{-1}(u_1), \Phi^{-1}(u_2)\big)}{\phi\big(\Phi^{-1}(u_1)\big)\phi\big(\Phi^{-1}(u_2)\big)} \text{ for } u_1, u_2 \in [0, 1].$$

Charpentier et al. (2006) and Geenens et al. (2017) used this technique in combination with kernel estimators of g. There is a variety of other techniques that try to estimate the copula density c directly: methods based on smoothing kernels (Gijbels and Mielniczuk 1990; Charpentier et al. 2006), Bernstein polynomials (Sancetta and Satchell 2004), B-splines (Kauermann et al. 2013), and wavelets (Genest et al. 2009). The method of Geenens et al. (2017) was used to produce the second row of Fig. 11.1.

Any of the above methods can be used for estimating pair copula densities in a vine copula model. The corresponding h-functions can be obtained by integrating the density estimate. With these two ingredients, the sequential estimation procedure of Sect. 7.2 can be adapted to yield a fully nonparametric estimate of the vine copula density. This approach was followed by Kauermann and Schellhase (2014) and Nagler and Czado (2016) with B-splines and kernels, respectively. Nagler et al. (2017) give an overview of the existing methods and compared them in a large simulation study. Overall, the local-likelihood kernel estimator of Geenens et al. (2017) seems to perform best, although spline methods may be preferable, if there is low dependence.

Nonparametric estimators of vine models also have appealing theoretical properties. Compared to parametric density estimators, nonparametric methods generally converge at slower rates. Even worse, the converge rate slows down as the number of variables increases. Put differently, the larger the number of variables, the larger is the number of observations required for reasonably accurate estimates. This phenomenon is widely known as the *curse of dimensionality* in nonparametric estimation (see e.g., Scott 2015).

Nagler and Czado (2016) proved that there is no curse of dimensionality for estimating simplified vine copula densities using nonparametric pair copula estimators. The simplifying assumption allows us to decompose the problem of estimating a d-dimensional density into $\binom{d}{2}$ bivariate estimation problems. As a result, the vine copula density estimator converges at a rate equivalent to a bivariate problem, no matter how large d is. This is a considerable advantage over other nonparametric techniques for multivariate density estimation. And even when the simplifying assumption is

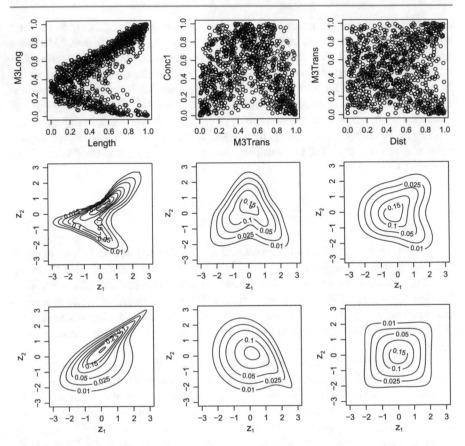

Fig. 11.1 MAGIC: Selected pairs from the MAGIC telescope data. First row: standardized rank plots, second row: empirical normalized contour plots, third row: contour plots of the best fitting parametric model

violated, this advantage makes vine copula based nonparametric estimators usually superior to traditional methods. This claim is supported by the simulation results contained in Nagler and Czado (2016). Hence, vine copulas with nonparametric pair copulas can be useful constructs in situations, where the dependence structure is not of primary interest.

Covariate-Dependent and Non-simplified Vine Copula Models

In some situations, it can be sensible to let the pair copulas in a vine depend on one or more covariates. For example, one may ask how the dependence between financial returns changes over time or how geographic characteristics influence the dependence between climatic variables. Another example is nonsimplified vines, where the copula parameter associated with edge $e = (a_e, b_e; D_e)$ in tree T_j for $j > 1$ may vary with the value of the conditioning variables $\boldsymbol{u}_{D_e}$.

More generally, suppose z_e is a p dimensional vector of covariates for edge e and the corresponding pair copula can be written as $c_{a_e,b_e;D_e}(\cdot,\cdot;z_e)$. If time is the only covariate, we define $z_e := t$ for all e. In a nonsimplified vine copula, we set $z_e := u_{D_e}$. The function $c_{a_e,b_e;D_e}(\cdot,\cdot;z_e)$ is the density of the copula associated with (U_{a_e}, U_{b_e}) where U_e depends on z_e. This dependence is difficult to specify in general. Therefore a common approach is to assume a parametric model for $c_{a_e,b_e;D_e}$ and let its parameter θ_e be a function of the covariates, i.e., $c_{a_e,b_e;D_e}(\cdot,\cdot) = c_{a_e,b_e;D_e}(\cdot,\cdot;\theta_e(z_e))$.

An early example of such a model is Erhardt et al. (2015b), who postulated a generalized linear model $\theta_e(z) = g(\beta_e^\top z)$ for the relationship between the dependence parameter and geographical dissimilarities. Similarly, a generalized linear model $\theta_e(u_{D_e}) = g(\beta_e^\top u_{D_e})$ can be used to relax the simplifying assumption. This was proposed by Han et al. (2017) in combination with a sparsity penalty, to automatically weed out unessential covariates.

The linearity assumption in the dependence parameter may still be too restrictive. To address this restriction Acar et al. (2012) proposed a semi-parametric model for a three-dimensional non-simplified vine copula. In the associated second tree, the dependence parameter θ_e is merely assumed to be a smooth function of the single covariate u_{D_e}. Acar et al. (2012) proposed to estimate this relationship by local-polynomial smoothing, but other smoothing methods like penalized splines can be employed as well. Vatter and Nagler (2018) used the latter technique to build a very general model, that allows for nonlinearity, nonsimplifiedness and an arbitrary number of covariates. More specifically, the pair copula parameters are assumed to follow the generalized additive model

$$\theta_e(z_e) = g\left(\sum_{k=1}^p s_{e,k}(z_{e,k})\right),$$

where $s_{e,k}$ are arbitrary smooth functions of a single component of $z_e \in \mathbb{R}^p$.

All models mentioned so far rely on the assumption that each pair copula belongs to a parametric family. A nonparametric approach for non-simplified vines was proposed by Schellhase and Spanhel (2018). The authors estimate conditional copulas with a single covariate using B-splines and use principal component analysis to reduce the covariate dimension in higher trees. A fully nonparametric method is yet to be explored.

11.2 Advances in Model Selection of Vine Copula Based Models

Finding Simplified Vine Tree Structures

Given copula data, one of the main tasks of the estimation of a simplified vine copula is finding a suitable tree structure. The tree structure determines on the one hand, which specific pair copulas are modeled, and on the other hand, which conditional independencies are implied by the simplifying assumption. The tree structure selection method proposed in Sect. 8.3, the so-called Dißmann's algorithm proposed in Dißmann et al. (2013), sequentially constructs the tree structure by maximizing dependence at each tree level. Here, dependence is measured in terms of absolute Kendall's τ values. However, Dißmann's algorithm does not account for the structure's implications on the simplifying assumption.

Kraus and Czado (2017b) develop two new algorithms, that select tree structures of vine copulas, which violate the simplifying assumption as little as possible. Using a recently developed test for constant conditional correlations (CCC) of Kurz and Spanhel (2017), their structure selection methods take information about the violation of the simplifying assumption into account. The first algorithm selects the first tree similar to Dißmann's algorithm, but additionally utilizes the p-values of the CCC tests as weights for the selection of higher trees. This yields tree structures, where the p-values of the CCC tests are rather large, implying that the simplifying assumption is not a severe restriction. The second algorithm of Kraus and Czado (2017b) proposes to fit a C-vine copula to the data, where for each tree level the root node is selected such that the occurrence of nonsimplifiedness in the next tree is minimized.

Kraus and Czado (2017b) demonstrate in a simulation study, that these algorithms are able to improve the model fit in terms of the Akaike information criterion (AIC) compared to Dißmann in many cases, especially when the dimension is large. Lastly, they revisit many data sets, that have already been studied in the vine copula context and thereby highlight the usefulness of their algorithms. For example, they show that the classic uranium data set of Cook and Johnson (1986), which is often used as an example of a data set for which the simplifying assumption is violated, can be suitably modeled as a simplified vine copula, when the tree structure is selected using their proposed algorithm.

Goodness-of-Fit Tests for Vine Copulas

Copula goodness-of-fit tests were introduced and discussed in Genest et al. (2006). More precisely, the hypothesis that the copula associated with the data belongs up to a parameter specified copula class is tested. Many different test statistics for this null hypothesis have been given, that follow two major approaches. In a first approach, the empirical copula process is considered, which is the difference between the fitted parametric copula and the empirical copula. These differences are then utilized to construct a multivariate Cramer–von Mises or multivariate Kolmogorov–Smirnov based test statistic. The other approach is based on first applying the Rosenblatt transformation (see Sect. 6.1) to transform the dependent components of the copula data to i.i.d components. These i.i.d components are then aggregated and standard univariate goodness-of-fit approaches are used. Their small sample performance was studied in Berg (2009) and Genest et al. (2009) mostly for bivariate copulas.

With the increased use of vine copulas especially in higher dimensions the need to construct and implement goodness-of-fit tests for the vine copula arose. Schepsmeier (2015, 2016) developed and investigated goodness-of-fit copula tests for the hypothesis that the underlying copula of a data set belongs to the class of a vine copula. Here, the vine tree structure as well as the pair copula families are assumed to be specified. For this, he first adapted the two approaches given above, but he also developed a goodness-of-fit statistics based on the information matrix equality and specification test proposed by White (1982). This extends the goodness-of-fit test for copulas introduced by Huang and Prokhorov (2014). While Schepsmeier (2015) concentrates on the performance of the tests, Schepsmeier (2016) develops the associated asymptotic theory. Schepsmeier (2014) additionally proposes an information ratio test, which is inspired by Zhou et al. (2012), and shows good small sample performance when compared to the alternatives. As is common in the area of copula of goodness-of-fit tests, there is no clear winner for a specific goodness-of-fit test in the context of vine copulas.

Model Distances Between Vine Copula Models

The two main advantages of vine copula models are, that they are extremely flexible and numerically tractable even in high dimensions. Being able to determine how much two specified models differ can be important in many cases. The Kullback–Leibler criterion of Kullback and Leibler (1951) is the most popular measure for this purpose and was introduced in Sect. 9.2. Often the criterion is also called the Kullback–Leibler distance, even though is does not satisfy the triangle inequality required for a distance. For example, Spanhel and Kurz (2015) try to find the simplified vine copula with the smallest Kullback–Leibler distance for a given copula. Other authors like Stöber et al. (2013) and Schepsmeier (2015) also use the Kullback–Leibler distance in the context of vine copulas. Unfortunately, the computation of the KL distance requires multivariate integration, which becomes numerically infeasible in high dimensions. To overcome this curse of dimensionality one could use Monte Carlo integration for approximating the true Kullback–Leibler distance. However, Monte Carlo methods are random and become imprecise and/or slow for higher dimensions.

Therefore, Killiches et al. (2018) develop alternative (nonrandom) methods based on the Kullback–Leibler distance. They use the relationship that the Kullback–Leibler distance can be written as the sum over expectations of Kullback–Leibler distances between univariate conditional densities. These expectations are then approximated by evaluation on specialized grids, that are chosen depending on the focus of the application. The resulting *approximate Kullback–Leibler distance* is simply an approximation of the true Kullback–Leibler distance and can be used from a computational point of view for up to five dimensions. The *diagonal Kullback–Leibler distance* concentrates on specific conditioning vectors, namely the ones on certain diagonals in the space. Therefore, it is not an approximation in the classical sense. However, the qualitative behavior is extremely similar to the one of the Kullback–Leibler distance. For up to ten dimensions this method can easily be used. When going to dimensions 30 and higher, the authors suggest reducing the number of evaluation points even further by considering only one principal diagonal, defining the *single diagonal Kullback–Leibler distance*. Many different examples and applications show, that the three proposed methods are suitable substitutes for the Kullback–Leibler distance. Further, they manage to outperform Monte Carlo integration from a computational point of view.

High-Dimensional Model Selection of Vine Based Models Using Graphical Structures

Vine copulas have been introduced as a flexible dependence modeling tool in the non-Gaussian case. However, exploring the properties of regular vines in the Gaussian case can help to draw important information for the vine tree structure selection of arbitrary regular vines and to obtain parsimonious models. More precisely, Gaussian models on *directed acyclic graphs (DAGs)*, also called *Bayesian Networks (BNs)* form a prominent class of graphical models for many applications in biology, psychology, ecology, and other areas. For example, Sachs et al. (2005) identified protein networks using this model class. In general, these are models for describing conditional independencies induced by a directed graph. Many theoretical results can be found in the books of Lauritzen (1996) or Koller and Friedman (2009).

To learn such a graphical model from data, most often the multivariate Gaussian distribution is assumed. In this case, either *score-based* or *constraint-based* methods are employed to learn the underlying DAG $\mathcal{G} = (V, E)$, where N is the set of nodes and E is a set of directed edges or *arrows* of the graph. Score-based approaches use a greedy procedure on the underlying data to find a graphical model with highest score such as the highest log-likelihood or lowest BIC. For constraint-based approaches, conditional independence tests are utilized to draw inference. For an implementation of this approach in R see Scutari (2010).

A first connection between Gaussian regular vine models and DAG's has been discovered by Brechmann and Joe (2014). For this, they consider truncated regular vines (compare to Remark 9.9) (Brechmann et al. 2012) and showed that truncated Gaussian vines can be represented as a special class of structural equation models (SEM) as for example discussed in Kaplan (2009). The special class of SEM's arising from truncated Gaussian vines of order k are models, where the nodes can be ordered and nodes of higher order can be expressed as a linear functions of exactly k nodes of lower order. Such a representation immediately leads to a DAG, where each node has at most k parents, i. e. at most k arrows pointing toward a node. A different characterization of k truncated regular vines is given in Haff et al. (2016).

Since there exist efficient structure learning algorithms for Gaussian DAG's in very high dimensions, it is interesting to investigate, which DAG's with at most k parents can be represented as k truncated Gaussian R-vine. This was the focus of Müller and Czado (2018a). Not every DAG with at most k parents can be represented as a k truncated Gaussian vine, but Müller and Czado (2018a) show that there exists a $k' \geq k$ such that the DAG with at most k parents can be represented as a k' truncated Gaussian vine. More precisely, the associated conditional independence relationships in the DAG can be represented by a k'-truncated R-vine. Since structure selection for Gaussian DAG's is much more faster than for arbitrary R-vines, it is attractable in high dimensions to first estimate a Gaussian DAG model, then find a representing R-vine tree structure and perform an additional maximum likelihood estimation of pair copula families and parameters on the chosen R-vine structure. Here we assume that an appropriate non-Gaussian vine tree structure is determined by the inclusion of edges which, indicate strong dependence between the (pseudo) copula data. These can be detected also by strong pairwise correlation between the (pseudo) copula data.

The benefit of this approach is, that given some sparse DAG model, i.e., with k low, a sparse R-vine representation with many independence copulas can be found. This speeds up the estimation process for R-vines copulas significantly.

Another approach using *undirected graphs* has been proposed recently by Müller and Czado (2017). The basic idea is similar to the previous approach, i. e. to first find a Gaussian graphical model, which can then be represented by an R-vine. However, this time undirected graphical models described by a *sequence of graphs* are considered. The associated sequence of graphs $\mathcal{G}_1, \ldots, \mathcal{G}_M$ are selected using the *graphical Lasso* of Friedman et al. (2008). This graph sequence induces a partition of the d dimensional data set into p_i different connected components for each element of the graph sequence $\mathcal{G}_i$, $i = 1, \ldots, M$. Furthermore, each of these graphs is again a graphical model. The idea is now to fit only an R-vine on each of the p_i connected components separately, and connect them afterwards. This reduces the complexity of an R-vine in d dimensions to a problem in at most η_i dimensions, where η_i is the size of the largest connected component in the graph $\mathcal{G}_i$. This also exploits (conditional) independence similar to the previous approach. Afterwards, the corresponding p_i R-vines are joined into one model. Additionally, the conditional independencies in the corresponding p_i sub-models can also be utilized to make models more parsimonious.

This approach also allows to improve the accuracy of the sequential greedy search algorithms introduced in Dißmann et al. (2013). Since many of the admissible edges for the vine tree sequence are set to the independence copula upfront by the conditional independencies encoded in the graphical models, replacing the value of an empirical Kendall's τ as weight by an actually calculated goodness-of-fit measure such as AIC or BIC for each edge becomes computationally tractable. Recall from Sect. 8.1 that the use of AIC or BIC for pair copula family selection requires the determination of AIC or BIC for each allowed family for each nonindependent pair copula term.

Selecting Vine Copulas in Ultra High Dimensions

The structure selection of Dißmann et al. (2013) presented in Sect. 8.3 has several drawbacks. At first, greedy search procedures do not insure globally optimal results in terms of goodness-of-fit measures as log-likelihood, AIC or BIC. This is especially a problem in higher dimensions. It is caused by the sequential bottom-up strategy employed for R-vine estimation. Since trees of lower order T_j put restrictions on higher order trees $T_{j\prime}$ for $j\prime > j$ by the proximity condition, suboptimal choices for lower order trees can have dramatic impact on higher order trees. Furthermore, it can also lead to excessive error propagation. Hence, greedy approaches for the structure selection have to be considered with caution if scaled to higher dimensions, say $d = 500$, because the errors propagate through too many tree levels.

An additional aspect in high-dimensional modeling is model parsimony. The total number of parameters in an R-vine model is $d(d-1)$ if for each edge a two-parameter pair copula family is used. This shows that models can become over-parametrized very easily in high dimensions. Thus, a significant reduction of the number of parameters is necessary. There are several ways to achieve this. First, testing each pair copula to be the independence copula is possible, however computationally inefficient, since each edge needs to be tested. Truncating the R-vine tree sequence $T_1, \ldots, T_{d-1}$ at some specific tree level k as suggested by Brechmann et al. (2012) demands the upfront definition of a sensible truncation level $k < d - 1$. Furthermore, these approaches rely on the principle of finding maximum spanning trees using empirical Kendall's τ for pair-wise data as weights, which might not be computationally feasible in hundreds or thousands of dimensions. Three different approaches have been introduced recently to handle these drawbacks.

First, Müller and Czado (2018a) link Gaussian DAGs to Gaussian R-vines. Thus, the DAG is used as a proxy for the R-vine tree structure to obtain a parsimonious dependence structure with many conditional independencies, and thus, independence pair copulas. Hence, we obtain a vine tree structure estimate, which is less prone to error propagation compared to Dißmann et al. (2013). This is the case, since we start with the entire vine tree structure from the beginning and do not need to sequentially select the vine tree structure. So only the problem of family selection and parameter estimation for each pair copula term is required. This task is simplified further by the use of the conditional independence information contained in the DAG model, which allows us to set many pair copulas to the independence copula upfront. However, since the overall approach resembles the method of Dißmann et al. (2013), the computational complexity is similar and hence, dependence models in more than $d = 500$ dimensions are difficult to estimate using this approach.

This motivates another method proposed in Müller and Czado (2018b), which does not use a connection between R-vines and graphical models in the Gaussian case, but to structural equation models (SEM) (see for example Kaplan 2009). The special class of SEM's considered are d ordered linear equations which encode the dependence in a d dimensional Gaussian distribution. Since these SEM's can be parametrized by correlation and partial correlation parameters, they represent under certain conditions a Gaussian R-vine as established in Brechmann and Joe (2014). This class of SEM's is described by linear equations for which parsimonious model selection in high dimensions can be performed using the *Least Absolute Shrinkage and Selection Operator (Lasso)* of Tibshirani (1996). It is shown that zero coefficients in the corresponding structural equations identify zero partial correlations and hence, independence pair copulas in a Gaussian R-vine. Thus, the Lasso can be used to perform model selection for Gaussian R-vines. This procedure takes also into account the proximity condition, however it is not based on finding spanning trees. Hence, the implementation is much faster than the previously introduced approaches and can scale up to about $d \sim 1000$ variables. Additionally, it determines a *regularization path* of the Gaussian R-vine, which allows to consider different levels of sparsity by only estimating one specific R-vine tree structure and different degrees of penalization.

The Lasso approach is already making a large step forward in terms of dimensions. However, it only reformulates the problem into d structural equations with at most $d - 1$ possible regressors on which the Lasso is applied to perform model selection. Going to even higher dimensions needs an additional change in the perspective, attained by a *divide-and-conquer* approach. Müller and Czado (2017) propose to consider not the original R-vine structure selection problem in d dimensions, but split it up in subproblems of dimension $d_T \ll d$. This means, a specific *threshold dimension* is introduced and the data is clustered to consist of subproblems of dimension of at most d_T. For this, again the graphical Lasso of Friedman et al. (2008) is used. Using these graphical models, we obtain a clustering into the desired subproblems, for which we can fit R-vines in significantly less time and with less complexity compared to the original dimension. Afterwards, these sub R-vines can be recombined. This leads to an overall parsimonious model with stronger within cluster dependence than the dependence between clusters. Employing this procedure, R-vine copulas can be estimated in $d > 2000$ dimensions in reasonable time and with tractable numbers of parameters.

More recently Nagler et al. (2018) propose a modified BIC criterion tailored to select sparse vine copulas in high dimensions. This criterion called *mBICV* can consistently distinguish between the true and alternative models under less stringent conditions than the classical BIC. Nagler et al. (2018) use mBICV to select the truncation level in a truncated vine copula or the threshold in a vine copula where pair copulas are set to the independence copula, when the associated dependence strength is less than the threshold. These sparse models are called *thresholded vines*. They illustrate their model selection approach for determining the Value at Risk (VAR) in a large stock portfolio.

11.3   Advances for Special Data Structures

Discrete Variables in Vine Copula Models

In many applications, we not only observe continuous variables, but also discrete variables. The uniqueness property of the copula is lost and interpretation is challenging as for example discussed in Genest and Nešlehová (2007). However, the copula is uniquely defined on the Cartesian product of the possible values of the components. This is sufficient for the practitioner and applications in low dimensions can for example be found in health sciences (Zimmer and Trivedi 2006), transportation (Spissu et al. 2009), insurance (Krämer et al. 2013) and finance (Koopman et al. 2015).

To determine the associated likelihood for these d dimensional copula models we need to evaluate rectangle probabilities. The usual inclusion/exclusion formula requires 2^d evaluations of the copula per observation, which is prohibitive in large dimensions. In contrast, Panagiotelis et al. (2012) show that a vine copula construction approach is feasible with only four evaluations per observation for each pair copula term, when all components are discrete. This is much more parsimonious than the previous approaches. In Stöber et al. (2015), this approach was extended to include both discrete and continuous components and Schallhorn et al. (2017)

show how this approach can be applied in a D-vine quantile approach, where either the response or some of the covariates are discrete. The last paper also contains a different nonparametric estimation approach by using kernel density estimates. In particular Nagler (2018) shows that *jittering* the discrete data as proposed earlier by Denuit and Lambert (2005) in an appropriate manner to make the data continuous, which does not destroy the asymptotic behavior of the kernel density estimator applied to the jittered data. This is in contrast to the case of a parametric copula model as discussed in Nikoloulopoulos (2013).

Vine Decompositions on Directed Acrylic Graphs

As already mentioned, *Gaussian directed acyclic graphs (DAG's)* or also called *Bayesian networks (BN)* play a prominent role for the selection of vine tree structures. They also have been extended to allow for non-Gaussian dependence by considering a similar decomposition of the joint distribution as in the vine distribution.

A prominent result for distributions on DAG's with global Markovian structure shows, that the associated joint distribution can be decomposed as the product of the distribution of each node given it parents (Lauritzen 1996). This factorization was first used by Hanea et al. (2006) in a Gaussian context and later extended by Bauer et al. (2012) to the non-Gaussian context to achieve a decomposition involving only bivariate copulas and univariate conditional distributions as arguments of the bivariate copula. They called their models *pair copula Bayesian networks (PCBN)*. While the decomposition is quite similar to the vine decomposition, a major difference is, that not always all required conditional distributions can be easily determined using only pair copula terms present in the decomposition. Therefore this might induce the need to determine those using higher dimensional integration based on the general recursion given in Theorem 4.10.

In a follow-up paper, Bauer and Czado (2016) derive general expressions for the copula decomposition allowing for non-Gaussian dependence for arbitrary DAG's. It also constructs a structure learning algorithm for non-Gaussian DAG's by extending the PC algorithm of Spirtes et al. (2000) to test for conditional independence under non-Gaussian dependence. For this a regular vine is fitted, such that the last tree T_{d-1} has as pair copula the copula between the two variables conditional on a set of variables to be tested. The test for the associated conditional independence is now facilitated by testing for the independence copula in the last vine tree T_{d-1}.

Another approach to allow for non-Gaussian dependence in a Bayesian network is to consider *nonparametric Bayesian networks (NPBN)*. These NPBN networks are discussed and reviewed in Hanea et al. (2015) and a recent extension to NPBN's in a dynamic setup using the decomposition of Bauer and Czado (2016) is given in Kosgodagan (2017).

D-vine Copula Based Quantile Regression

Predicting conditional quantiles, also known as *quantile regression*, are important in statistical modeling and especially desired in financial applications. Risk managers use this approach for portfolio optimization, asset pricing and the evaluation of systemic risk. The standard methodology is linear quantile regression (Koenker and

Bassett 1978), which can be seen as an extension of ordinary least squares estimation. It has disadvantages such as the assumption of Gaussianity and the occurrence of quantile crossings in data applications. When margins are Gaussian, the only dependence structure, which has linear conditional quantiles, is the Gaussian dependence as shown in Bernard and Czado (2015), thus limiting its applicability to financial applications.

In Kraus and Czado (2017a) a new semi-parametric quantile regression method is introduced, which is based on sequentially fitting a conditional likelihood optimal D-vine copula to given regression data Y with covariates $x = (x_1, \ldots, x_p)'$. This results in highly flexible models with easily extractable conditional quantiles. We note, that the quantiles do not cross for any level by construction.

The main contribution of Kraus and Czado (2017a) is their proposed algorithm, which sequentially selects the set of important covariates $u = (u_1, \ldots, u_K)'$ on the copula scale using estimated marginal distributions. It estimates a D-vine copula $C(v, u_1, \ldots, u_p)$ for the copula response V and the random covariates $U := (U_1, \ldots, U_p)'$ with many desirable features. The resulting conditional distribution of V given $U = u$ is called the *vine copula regression model*. The algorithm adds a covariate U_{k+1} to the current vine copula regression model, if the conditional likelihood of V given $U_c = (u_1, \ldots, u_k)'$ and $U_{k+1} = u_{k+1}$ is increased over the one, which only includes U_c. This means that the algorithm stops adding covariates to the copula regression model, when the addition of any of the remaining covariates would not improve the model's conditional likelihood significantly. This gives an automatic forward selection of the covariates.

As a result, the chosen models are parsimonious and at the same time flexible in their choice of required pair copulas. The special nature of D-vine copulas allows expressing the conditional distribution of a leaf in the first tree T_1 given all other variables as a convolution of associated h functions. Thus, quantiles can be determined by the appropriate convolution of inverse h functions. This property was already exploited in the simulation algorithm for D-vine copulas presented in Sect. 6.4. It further allows for efficient determination of the conditional copula quantile for any level, if the univariate inverse h functions are available analytically. Otherwise, only numerical inversion of one-dimensional functions are required. To get the associated conditional quantiles on the original scale of the response variable Y, an estimate of the quantile function of Y is required.

In summary, for these models, the estimated conditional quantiles may strongly deviate from linearity and estimated quantiles do not cross for different levels. Further, the classical issues of quantile regression such as dealing with transformations, interactions and collinearity of covariates are automatically taken care of.

Kraus and Czado (2017a) demonstrate in their simulation study the improved accuracy and decreased computation time of their approach compared to already established nonlinear quantile regression methods, such as boosted additive quantile regression (Fenske et al. 2012) or nonparametric quantile regression (Li et al. 2013). Finally, they underline the usefulness of their proposed method in an extensive financial application to international credit default swap data including stress testing and value at risk prediction.

After the publication of Kraus and Czado (2017a), further research on the subject followed. On the theoretical side, Rémillard et al. (2017) investigated the asymptotic limiting distribution of the conditional quantile estimator. Further, Schallhorn et al. (2017) extended the methodology to allow also for mixed discrete and continuous data to be analyzed. On the application side, Fischer et al. (2017) used D-vine quantile regression to do an in-depth analysis of the German economy with regards to stress testing.

The R package `vinereg` of Nagler and Kraus (2017) implements the D-vine quantile regression for continuous and mixed discrete/continuous data.

Vine Factor Copulas

Factor analysis is a widely used technique for sparse modeling in statistics (see for example Comrey and Lee 2013). The variability of variables is described by underlying unobserved latent factors. Often it is assumed, that the variables follow a normal distribution. Krupskii and Joe (2013) propose a *factor structure for vine copulas*. This model requires, that variables are independent given latent factors. More precisely, we consider d uniform $(0, 1)$ distributed variables $U_1, \ldots U_d$ and p i.i.d. uniform $(0, 1)$ distributed latent factors $W_1, \ldots W_p$. In the factor copula model with p latent factors we assume, that the conditional distribution of

U_i and U_j given $W_1, \ldots W_p$ are independent for $i \neq j, i, j \in \{1, \ldots, d\}$.

This implies that the joint density c of $U_1, \ldots U_d$, can be written as

$$c(u_1, \ldots u_d) = \int_{[0,1]^p} \prod_{j=1}^d c_{U_j|W_1,\ldots,W_p}(u_j|w_1, \ldots, w_p)dw_1 \cdots dw_p.$$

As we see, we need to model the dependence between one of the variables U_j and the factors $W_1, \ldots, W_p$. If p is small this lead to a significant reduction in model complexity.

Considering the above model with one factor, i.e $p = 1$, we only need to model d bivariate copulas to model the dependence. In contrast a full R-vine in d dimensions requires $\frac{d(d-1)}{2}$ bivariate copulas. Krupskii and Joe (2013) show how model parameters can be estimated using maximum likelihood estimation. Furthermore, they show how certain properties, like tail dependence, of the conditional distributions $C_{U_1|W_1,\ldots,W_p}, \ldots, C_{U_d|W_1,\ldots,W_p}$, for $p = 1, 2$, influence the distribution of $U_1, \ldots U_d$.

Schamberger et al. (2017) provide an alternative to maximum likelihood estimation. They analyze a one-factor vine copula model in a Bayesian framework and show how parameters are estimated using Markov Chain Monte Carlo (MCMC). An advantage of this method is, that the posterior distribution of the latent factors $W_1, \ldots W_p$ is estimated in contrast to the frequentist approach proposed by Krupskii and Joe (2013), where integration over the factor variables is used. The estimated posterior distribution can yield useful information about the latent factors. Schamberger et al. (2017) also provide an application where factor copulas are used to analyze portfolio returns. As already mentioned Kreuzer and Czado (2018) use a different MCMC algorithm based on Hamiltonian Monte Carlo for the Bayesian analysis of one-factor vine copula model allowing for stochastic volatility models for the margins. A joint estimation approach of marginal and copula parameters was taken and the forecasting performance of a portfolio evaluated.

Since vine factor copulas is an attractive model class further extensions have been studied.

Nikouloulopoulos and Joe (2015) show, how the vine factor copula model can be applied to item response data, i.e. the variables considered are on an ordinal scale and not continuous as in most applications. This allows them to analyze data from surveys.

In the factor copula model each latent factor is linked to each of the variables $U_1, \ldots U_d$. Krupskii and Joe (2015a) provide an extension of this, where the variables are split into different groups and latent factors are linked to certain groups. They demonstrate how this model can be applied to estimate the value at risk and the conditional tail expectation for financial returns. The groups in their model correspond to different industry sectors. Recently, the factor vine copula approach has been extended to allow for spatial structures in Krupskii et al. (2018).

Somewhat different approaches to combine factor structures with copulas was followed by Oh and Patton (2017) and Ivanov et al. (2017).

Lee and Joe (2018) provide multivariate extreme value models with factor structure. They derive the tail dependence functions of the vine factor copula model and use them to construct new models. In general, extreme value theory is not easily established for vine copula models. A first illustration in three dimensions is given in Killiches and Czado (2015).

11.4 Applications of Vine Copulas in Financial Econometrics

The dominant application area of copula-based statistical models is in the area of finance. Here the need to forecast the value of risk for very high levels for a risky asset, which is dependent on other assets, is of primary interest. So models must allow for tail dependence such as vines (Joe et al. 2010). The availability of daily data for a multitude of financial assets fosters the application of copula models. Thus also vine copula models play an important role in this context. A first review of vine copula based applications to financial data is given in Aas (2016). It includes references for applications to market risk, capital asset pricing, credit risk, operational risk, liquidity risk, systemic risk, portfolio optimization, option pricing. In the following, we concentrate on two areas, which are of current research interest for financial data.

Vine Copulas with Time-Varying Dependence Structures

The availability of daily data also allows to investigate the dependency over time. Simple approaches include rolling window approaches, where static dependence is assumed within a time window. While this gives first insights, whether a dependence structure is changing over time, more sophisticated models are needed.

A first approach to allow for changing dependence structures is to assume a regime switching approach. In the context of the vine copulas this was first proposed and investigated by Chollete et al. (2009) using C-vines, while Stöber and Czado (2014) used general R-vines. Recent applications are Fink et al. (2017) and Gurgul and Machno (2016). In all of these approaches, the regime switching only takes place on the copula scale. The case where margins are also switching is considered in Stöber (2013), however, the models require many additional parameters.

A more general approach is to allow the copula parameter to vary with time. The time variation can be induced by covariates such as lagged observed responses or other time-varying quantities or allowing for a stochastic time dynamic. The first class are examples of observation driven models and are surveyed in Manner and Reznikova (2012), which are much simpler to estimate. Another more recent semi-parametric application observation driven model is given in Vatter and Chavez-Demoulin (2015). However, the interpretation of parameter estimates in observation driven models is limited to the data set at hand, while the models with a stochastic time dynamic allow for data independent interpretation.

Dynamic parameter-driven models were studied in a bivariate copula context for example in Almeida and Czado (2012) using a Bayesian approach for estimation, while Hafner and Manner (2012) used an efficient importance sample approach for estimation. The extension of this approach to higher dimensional D-vine copulas was studied in Almeida et al. (2016). This approach was also used for modeling the joint distribution of electricity prices in Manner et al. (2019).

Vines for Modeling Time-Series of Realized Volatility Matrices

Given the increasing availability of high-frequency data volatility and covariance, modeling and forecasting have become of particular interest in financial econometrics. Reliable models, which are flexible enough, to account both for the continuously tightening interactions and interconnectedness between financial markets are needed.

The sum of squared intraday returns constitutes a consistent estimate of ex-post realized volatility and realized covariances. This makes naturally latent variables observable and measurable (see for example Doléans-Dade and Meyer 1970 and Jacod 1994). Thus Barndorff-Nielsen and Shephard (2004) show that standard techniques can be used for time-series modeling of realized covariance matrices. By doing so, algebraic restrictions such as positive semi-definiteness and symmetry of the forecasts have to be satisfied.

One of the most frequent modeling approaches to satisfy the restrictions is based on data transformation. For example, Bauer and Vorkink (2011) use the matrix logarithm function or Andersen et al. (2003) utilize the Cholesky decomposition applied to the time-series of realized covariance matrices. Chiriac and Voev (2011) model the so-obtained series of Cholesky elements using a vector ARFIMA process. In contrast, Brechmann et al. (2018) develop regular vine copula models to further explore the specific dependencies among the Cholesky series induced by this nonlinear data transformation. They are able to improve the Cholesky decomposition based model approach. A study on value at risk forecasting gives evidence, that the prediction model incorporating regular vine models performs best due to more accurate forecasts.

However, regular vines do not only prove themselves useful as an ingredient within the Cholesky decomposition based model approach, but allow for an alternative and novel data transformation themselves. Using partial correlation vines a joint prediction model of realized variances and realized correlations is proposed. Starting point are the univariate time-series of realized partial correlations, which can be obtained via recursive computation given in Theorem 2.14. By choosing a regular vine structure $\mathcal{V}$ on d elements, a subset of $d(d-1)/2$ pairwise standard correlations and partial correlations is selected. Here, for an edge $e \in \mathcal{V}$ the partial correlation $\rho_{i,j;D_e}$ coincides with the conditional constraint $(i, j|D_e)$ associated with edge e $(i, j \in \{1, \ldots, d\}, i \neq j, D_e \subset \{1, \ldots, d\} \setminus \{i, j\})$. Recall that Bedford and Cooke (2002) show, that there is a bijection between the (partial) correlations specified by any regular vine structure and the set of symmetric and positive definite correlation matrices. Also, any partial correlation vine specifies algebraically independent (partial) correlations, i.e., positive definiteness is automatically guaranteed. To the series of realized (partial) correlations specified by the selected regular vine structure elaborate time-series models such as heterogeneous autoregressive (HAR) processes can be applied to account for, e.g., long-memory behavior and multifractal scaling. Extensions by considering GARCH augmentations are discussed in Corsi et al. (2008). They allow to model non-Gaussianity and volatility clustering in the marginal time-series. Skewed error distributions for the residuals capture possible high skewness and kurtosis and have been considered in Bai et al. (2003) and Fernández and Steel (1998). Regular vine structure selection on the copula data arising from the marginal time-series modeling can be applied using the top-down Dißmann algorithm of Sect. 8.3. In contrast to the standard Dißmann algorithm the considered edge weights exclusively rely on historical information contained in the (partial) correlation time-series and thus allow for a flexible and dynamic parameterization over time.

Barthel et al. (2018) introduce the partial correlation vine approach in detail and explore its applicability based on data from the NYSE TAQ database compared to Cholesky decomposition based benchmark models.

11.5 Applications of Vine Copulas in the Life Sciences

Vine copulas have also found applications in the life sciences. For example, Nikoloulopoulos (2017) evaluates the diagnostic test accuracy using a trivariate vine model for the number of true positives (diseased person correctly diagnosed), true negatives (healthy person correctly diagnosed) and diseased persons, while Diao and Cook (2014) consider a multiple multistate model with right censoring. It reduces to a four-dimensional D-vine in the case, where there are two progressive states for each of two components.

As we already mentioned vines can also handle discrete structures, which do often occur in the life science. A first application was the modeling of the dependence structure among binary indicators of chronic conditions and the body mass index in the elderly is given in Stöber et al. (2015). Schmidl et al. (2013) show that a MCMC based Bayesian analysis of biological dynamical systems can profit from using a vine dependence model to speed up the convergence to the posterior distribution.

We now outline advances vine based models can make to the analysis of multivariate survival analysis and to longitudinal data analysis.

Vine Copulas for Modeling Dependence in Right-Censored Event Time Data

In biomedical, engineering or actuarial studies primary focus lies on the time until a predefined event occurs. However, due to a limited study horizon or due to early drop-out the event of interest might not be recorded for all observed individuals, but only a so-called right-censoring time might be registered. The resulting lack of information has to be carefully taken into account by inference tools applied to right-censored data in order to arrive at a sound statistical analysis.

Often, multivariate event time data are generated from clusters of equal size and hence exhibit possibly complex association patterns, which require elaborate dependence models. Barthel (2015) and Barthel et al. (2018) extend the flexible class of *vine copula models to right-censored event time data*. To trivariate and quadruple data they apply a two-stage estimation approach. In the first step they consider both parametric and nonparametric approaches for modeling the marginal survival functions. In the second step dependence of the associated pseudo data are modeled parametrically using maximum likelihood. A likelihood expression in terms of vine copula components adapted to the presence of right censoring is established. It is shown that due to the construction principle of vine copulas and right-censoring single and double integrals shop up requiring numerical integration for likelihood evaluation. A sequential estimation approach is proposed and implemented to facilitate the computationally challenging optimization problem. A detailed investigation of the mastitis data of Laevens et al. (1997), where primiparous cows (the clusters) are observed for the time to mastitis infection in the four udder quarters, stresses the need for elaborate copula models to reliably capture the inherent dependence.

Alternatively, an event might be recurrent for each sample unit. For example, for children in a medical study a series of subsequent asthma attacks might be recorded. Typically, the times between subsequent asthma attacks, the so-called gap times, are dependent on each child. Further, some children are more prone to asthma attacks than others such that different number of gap times are recorded. Another challenge arises from the recurrent nature of the data, which results in gap times subject to induced dependent right censoring. Barthel et al. (2017) investigate the class of D-vine copula models, which can easily handle the unbalanced data setting and which naturally capture the serial dependence inherent in recurrent event time data.

Modeling Longitudinal Data with D-Vine Copulas

Analyses of longitudinal studies go back the nineteenth century, where astronomers like Airy (1861) investigated repeated measurements of celestial phenomena. Today there exists extensive literature on statistical concepts and applications of longitudinal data. For a review see Fitzmaurice et al. (2008) or Diggle and Donnelly (1989).

Linear mixed models extend the classical linear models by combining the fixed effects with individual-specific random effects. Diggle (2002) and Verbeke and Molenberghs (2009) provide thorough introductions to the class of linear mixed models. These models are extremely popular, since they can easily be designed, estimated, interpreted and applied. However, independently from the actual specification, the resulting model remains Gaussian by definition. This implies in particular that the underlying dependence structure is always Gaussian.

As dependence modeling became more and more popular in all areas of applications, copulas were also applied for modeling repeated measurement data. Meester and MacKay (1994) were the first one to create a model for bivariate clustered categorical data. Other references are for example Lambert and Vandenhende (2002), Shen and Weissfeld (2006) and Sun et al. (2008).

Since D-vine copulas directly imply a serial ordering, they are particularly suited for modeling temporal dependence. Further, they are flexible and sub-models which are needed for conditional prediction are analytically available, making these models very attractive for application. Therefore, authors like Smith et al. (2010) use them to model (univariate) longitudinal data. In Smith (2015) and Nai Ruscone and Osmetti (2017) multivariate longitudinal data are considered. Semicontinuous longitudinal insurance claims are analyzed in Shi and Yang (2018) with the help of a mixed D-vine copula. However, all the above references only work in the so-called balanced setting, where all individuals must have the same number of measurements. Shi et al. (2016) use a Gaussian copula in the unbalanced setting.

Killiches and Czado (2018) present an approach, that uses a D-vine copula based model without restrictions on the margins for modeling longitudinal data in the unbalanced setting. The model can be used for understanding the association between the measurements of one individual. Further, conditional prediction of future events is possible. For this purpose the authors use the fact that the necessary conditional quantiles are explicitly given. For model selection an adjusted version of the *Bayesian information criterion* (BIC) is developed in the unbalanced setting. The model is compared to linear mixed models and found to be an extension of a large class of LMMs. Further, a sequential maximum likelihood based estimation method for the D-vine copula is proposed, which resembles the approach of Dißmann et al. (2013) for complete data. A simulation study shows good performance of the estimator in small samples. In an application to medical data both linear mixed models and D-vine copula based models are fitted. The quality of the fitted models is compared based on the derived model selection criteria and the performance of conditional quantile prediction. Barthel et al. (2017) extend the concept of Killiches and Czado (2018) to right-censored recurrent event data.

11.6   Application of Vine Copulas in Insurance

Applications of copulas in insurance started with the bivariate modeling of the number of claims and the average claim size to assess the total loss (Czado et al. 2012 and Krämer et al. 2013). In Erhardt and Czado (2012) vine copulas were used in modeling different types of health insurance claims over several time periods. Special care had to be taken to allow for zero inflation.

Shi and Yang (2018) use a pair copula construction framework for modeling semicontinuous longitudinal claims. In the proposed framework, a two-component mixture regression is employed to accommodate the zero inflation and thick tails in the claim distribution. The temporal dependence among repeated observations is modeled using a sequence of bivariate conditional copulas based on a D-vine.

Timmer et al. (2018) study the asset allocation decision of a life insurance company using a vine copula approach.

11.7   Application of Vine Copulas in the Earth Sciences

Spatial data is characterized by the distance between two observed locations. In Gräler and Pebesma (2011) pairwise distances are arranged in a set of distance classes. The associated spatial data within one class is used to fit a bivariate copula. These class dependent copulas are joined by convex interpolation to form a bivariate copula for arbitrary pairwise distances. To predict the median value at an unobserved location s_0 Gräler and Pebesma (2011) uses the observed values of d nearest neighbors $x_1, \ldots, x_d$ with locations $s_1, \ldots, s_d$ to construct a $d + 1$ dimensional C-vine truncated after the first tree, where the root in the first tree node is given by the forecast X_0 at the unobserved location s_0. The pair copula between the neighbor X_j and X_0 is chosen as the fitted distance based copula with distance d_{j0} between the locations s_j and s_0. Now the conditional distribution of X_0 given the observed values $x_1, \ldots, x_d$ is determined by d dimensional integration and using a fitted marginal distribution based on all observed data. This local neighborhood model is estimated using a composite likelihood approach and applied to model the spatial distribution of heavy metal concentration at the Meuse river. In a follow-up paper Gräler (2014) applied this local approach to fit a non-Gaussian skewed spatial model to an emergency scenario data set from the Spatial Interpolation Comparison in 2014, where it performed very well in the competition. Again a composite likelihood inference approach was followed using four-dimensional C-vines in Erhardt et al. (2015b). For prediction, a different conditional distribution was utilized.

A completely different approach to model the spatial dependency was used in Erhardt et al. (2015a). Here the dependency was captured through distance-based regression effects on the parameter of regular vines. This allows for a parsimonious model and was successfully applied to model German temperature data.

In hydrology often spatial pattern also play a major role. For the simulation of precipitation dependence pattern Hobæk Haff et al. (2015) used the pair copula construction, while Bevacqua et al. (2017) used vines to construct multivariate statistical models for compound events in floods. Time-series of ocean wave heights and mean zero-crossing periods were studied in Jäger and Nápoles (2017). For this they modified the approach of Brechmann and Czado (2015). This approach allows to incorporate cross serial dependence between time-series. For multi site stream flow simulation Pereira et al. (2016) followed the approach of Erhardt et al. (2015a) to model the spatial dependence among the sites. In a follow up paper (Pereira et al. 2017) they allowed the inclusion of periodic effects. Vines were also used to construct multivariate drought indices (Erhardt and Czado 2018).

In weather forecasting often ensembles are generated from numerical weather prediction (NWP) models, consisting of a set of differential equations describing the dynamics of the atmosphere. It is common to use statistical methods to post process these ensembles. Möller et al. (2018) a new post-processing method based on D-vine quantile regression is developed and shown to have superior performance over benchmarks for longer forecast horizons.

11.8   Application of Vine Copulas in Engineering

With the availability of simulated or real data, dependence models are also applied in engineering. For example Schepsmeier and Czado (2016) considered data from crash simulation experiments. Vine models were used to characterize the dependence. Another example is Höhndorf et al. (2017), who investigated the relationship among variables arising from operational flight data using marginal regression models together with a vine copula.

Vine copula based models have also been successfully used in reliability analysis for mechanical structures. Jiang et al. (2015) use a vine copula to formulate a reliability model to quantify the uncertainties in loadings, material properties, structure sizes etc. for mechanical structure. Here, dependencies among derived stochastic variables are modeled, which are visible in real data sets and thus allowing for more realistic reliability models. Hu and Mahadevan (2017) capture additional time effects in a reliability analysis.

11.9   Software for Vine Copula Modeling

Freely available software played a major role in the success of vine copula models. The algorithms for inference and simulation are quite complex and their implementation requires effort and expertise. Applied researchers were relieved of that burden early on by the R package CDVine of Brechmann and Czado (2013), first released in May 2011. Reflecting the (then) current state of research, the package was limited to C- and D-vines.

Today the most widely used package is `VineCopula` provided by Schepsmeier et al. (2018), which is the successor of `CDVine`. It allows for arbitrary vine structures and brings with it additional copula families and a lot of extra functionality for modeling dependence with bivariate or vine copulas. The package's functionality covers most of the material in this book, including

- statistical functions (densities, distributions, simulation),
- inference algorithms (parameter estimation, model selection),
- tools for exploratory data analysis and visualization.

The functions used in this book are summarized in Table 11.1.

A recent alternative is the `vinecopulib` project (www.vinecopulib.org). Its core is an efficient `C++` implementation of the most important features of `VineCopula`, which is especially useful for high-dimensional applications. In addition, it allows to mix parametric with nonparametric pair copulas and provides interfaces to both R (Nagler and Vatter 2017) and `Python` (Arabas et al. 2017), respectively. Beyond that, there is a `MATLAB` toolbox for vine copulas with an associated `C++` library (Kurz 2015).

There are also a few more specialized R packages related to vine copula models:

- `CDVineCopulaConditional` (Bevacqua 2017) provides conditional sampling algorithms for C- and D-vines,
- `gamCopula` (Vatter and Nagler 2017) allows the parameters of vine copulas to vary flexibly with covariates,
- `kdevine` (Nagler 2017b), `penRvine` (Schellhase 2017b), and `pencopula Cond` (Schellhase 2017a) for nonparametric estimation of vine copulas,
- `pacotest` (Kurz 2017) implements a test for the simplifying assumption,
- `vinereg` (Nagler and Kraus 2017) implements the D-vine quantile regression of Kraus and Czado (2017a) and Schallhorn et al. (2017).

Table 11.1 Software: Functions of the VineCopula package used in this book

	Sections	Function	Short description
Chapter 3	3.1–3.4, 3.6	BiCop	Bivariate copula families
		BiCopPDF	Computes the copula density
		BiCopCDF	Computes the copula distribution function
		BiCopHfunc	Computes h-functions
	3.5	BiCopPar2Tau	Converts copula parameter to Kendall's τ
		BiCopTau2Par	Converts Kendall's τ to copula parameter
	3.7	BiCopPar2TailDep	Converts copula parameter to tail dependence coefficients
	3.8	pobs	Computes standardized ranks of data
		plot.BiCop	Plots a 3D-surface of copula densities
		contour.BiCop	Plots (normalized) contours of copula densities
		BiCopKDE	Plots empirical (normalized) contours
	3.9	BiCopSim	Simulates from a bivariate copula
Chapter 5	5.3–5.5	RVineMatrix	Constructs vine copula from structure-, family-, and parameter- matrices
		RVinePDF	Computes density of vine copula model
	5.2	plot.RVineMatrix	Plots the vine tree sequence
Chapter 6	6.2–6.5	RVineSim	Simulates from a vine copula
Chapter 7	7.1	RVineLogLik	Computes log-likelihood of a vine copula
		RVineMLE	Computes joint MLE for fixed vine structure and families
	7.2	RVineSeqEst	Computes sequential MLE for fixed vine structure and families
	7.3	RVineHessian	Computes Hessian matrix
		RVineStdError	Computes standard errors

(continued)

Table 11.1 (continued)

	Sections	Function	Short description
Chapter 8	8.1	`RVineCopSelect`	Selects families + estimates parameters for fixed structure
	8.2	`RVineStructureSelect`	Selects families and structure + estimates parameters
Chapter 9	9.1	`RVineAIC/RVineBIC`	Computes AIC and BIC of vine copula
	9.3	`RVineVuongTest`	Computes test statistic and p-value of Vuong test
	???	`RVineClarkeTest`	Computes test statistic and p-value of Clarke test

References

Aas, K. (2016). Pair-copula constructions for financial applications: A review. *Econometrics*, *4*(4), 43.

Aas, K., Czado, C., Frigessi, A., & Bakken, H. (2009). Pair-copula constructions of multiple dependence. *Insurance, Mathematics and Economics*, *44*, 182–198.

Acar, E. F., Genest, C., & Nešlehová, J. (2012). Beyond simplified pair-copula constructions. *Journal of Multivariate Analysis*, *110*, 74–90.

Acar, E. F., Genest, C., & Nešlehová, J. (2012). Beyond simplified pair-copula constructions. *Journal of Multivariate Analysis*, *110*, 74–90.

Airy, G. (1861). *On the algebraic and numerical theory of errors of observations and the combination of observations*. London: Macmillan.

Akaike, H. (1973). Information theory and an extension of the maximum likelihood principle. In B. N. Petrov & F. Csaki (Eds.), *Proceedings of the Second International Symposium on Information Theory Budapest* (pp. 267–281). Akademiai Kiado.

Akaike, H. (1998). Information theory and an extension of the maximum likelihood principle. *Selected Papers of Hirotugu Akaike* (pp. 199–213). Berlin: Springer.

Almeida, C., & Czado, C. (2012). Efficient Bayesian inference for stochastic time-varying copula models. *Computational Statistics and Data Analysis*, *56*(6), 1511–1527.

Almeida, C., Czado, C., & Manner, H. (2016). Modeling high-dimensional time-varying dependence using dynamic D-vine models. *Applied Stochastic Models in Business and Industry*, *32*(5), 621–638.

Andersen, T. G., Bollerslev, T., Diebold, F. X., & Labys, P. (2003). Modeling and forecasting realized volatility. *Econometrica*, *71*(2), 579–625.

Anderson, T. W. (1958). *An introduction to multivariate statistical analysis*. Hoboken, NJ, Wiley & Sons.

Arabas, S., Nagler, T., & Vatter, T. (2017). *Pyvinecopulib: high performance algorithms for vine copula modeling*. Python library.

Azzalini, A., & Capitanio, A. (2014). *The skew-normal and related families.*, Institute of mathematical statistics monographs Cambridge: Cambridge University Press.

Baba, K., Shibata, R., & Sibuya, M. (2004). Partial correlation and conditional correlation as measures of conditional independence. *Australian and New Zealand Journal of Statistics*, *46*(4), 657–664.

© Springer Nature Switzerland AG 2019
C. Czado, *Analyzing Dependent Data with Vine Copulas*, Lecture Notes
in Statistics 222, https://doi.org/10.1007/978-3-030-13785-4

Bai, X., Russell, J. R., & Tiao, G. C. (2003). Kurtosis of GARCH and stochastic volatility models with non-normal innovations. *Journal of Econometrics, 114*(2), 349–360.

Barndorff-Nielsen, O. E., & Shephard, N. (2004). Econometric analysis of realized covariation: High frequency based covariance, regression, and correlation in financial economics. *Econometrica, 72*(3), 885–925.

Barthel, N. (2015). Multivariate Survival Analysis using Vine-Copulas. Master's thesis, Technische Universität München.

Barthel, N., Czado, C., & Okhrin, Y. (2018). A partial correlation vine based approach for modeling and forecasting multivariate volatility time-series. arXiv:1802.09585.

Barthel, N., Geerdens, C., Czado, C., & Janssen, P. (2017). Modeling recurrent event times subject to right-censoring with D-vine copulas. arXiv:1712.05845.

Barthel, N., Geerdens, C., Killiches, M., Janssen, P., & Czado, C. (2018). Vine copula based likelihood estimation of dependence patterns in multivariate event time data. *Computational Statistics and Data Analysis, 117*, 109–127.

Bauer, A., & Czado, C. (2016). Pair-copula Bayesian networks. *Journal of Computational and Graphical Statistics, 25*(4), 1248–1271.

Bauer, A., Czado, C., & Klein, T. (2012). Pair-copula constructions for non-Gaussian DAG models. *Canadian Journal of Statistics, 40*(1), 86–109.

Bauer, G. H., & Vorkink, K. (2011). Forecasting multivariate realized stock market volatility. *Journal of Econometrics, 160*(1), 93–101.

Bedford, T., & Cooke, R. M. (2001). Monte Carlo simulation of vine dependent random variables for applications in uncertainty analysis. In *Proceedings of ESREL2001*. Italy: Turin.

Bedford, T., & Cooke, R. M. (2002). Vines: A new graphical model for dependent random variables. *Annals of Statistics, 30*(4), 1031–1068.

Beirlant, J., Goegebeur, Y., Segers, J., & Teugels, J. (2006). *Statistics of extremes: theory and applications*. New York: Wiley.

Berg, D. (2009). Copula goodness-of-fit testing: an overview and power comparison. *The European Journal of Finance, 15*(7–8), 675–701.

Bergsma, W. (2011). Nonparametric testing of conditional independence by means of the partial copula. arXiv:1101.4607.

Bergsma, W. P. (2004). *Testing conditional independence for continuous random variables*. Eindhoven: Eurandom.

Bernard, C., & Czado, C. (2015). Conditional quantiles and tail dependence. *Journal of Multivariate Analysis, 138*, 104–126.

Bevacqua, E. (2017). CDVineCopulaConditional: Sampling from Conditional C- and D-Vine Copulas. *R package version 0.1.0*.

Bevacqua, E., Maraun, D., Haff, I. H., Widmann, M., & Vrac, M. (2017). Multivariate statistical modelling of compound events via pair-copula constructions: analysis of floods in ravenna (italy). *Hydrology and Earth System Sciences, 21*(6), 2701.

Brechmann, E. (2010). Truncated and simplified regular vines and their applications. Master's thesis, Technische Universität München.

Brechmann, E. C., & Czado, C. (2013). Risk management with high-dimensional vine copulas: An analysis of the Euro Stoxx 50. *Statistics and Risk Modeling, 30*(4), 307–342.

Brechmann, E. C., & Czado, C. (2015). COPAR-multivariate time series modeling using the copula autoregressive model. *Applied Stochastic Models in Business and Industry, 31*(4), 495–514.

Brechmann, E. C., Czado, C., & Aas, K. (2012). Truncated regular vines and their applications. *Canadian Journal of Statistics, 40*(1), 68–85.

Brechmann, E. C., Heiden, M., & Okhrin, Y. (2018). A multivariate volatility vine copula model. *Econometric Reviews, 37*(4), 281–308.

Brechmann, E. C., & Joe, H. (2014). Parsimonious parameterization of correlation matrices using truncated vines and factor analysis. *Computational Statistics and Data Analysis, 77*, 233–251.

Charpentier, A., Fermanian, J.-D., & Scaillet, O. (2006). The estimation of copulas: Theory and practice. In J. rn Rank (Ed.), *Copulas: From Theory to Application in Finance*. Risk Books.

Chiriac, R., & Voev, V. (2011). Modelling and forecasting multivariate realized volatility. *Journal of Applied Econometrics, 26*(6), 922–947.

Chollete, L., Heinen, A., & Valdesogo, A. (2009). Modeling international financial returns with a multivariate regime-switching copula. *Journal of Financial Econometrics, 7*(4), 437–480.

Clarke, K. A. (2007). A simple distribution-free test for nonnested model selection. *Political Analysis, 15*(3), 347–363.

Coles, S., Bawa, J., Trenner, L., & Dorazio, P. (2001). *An introduction to statistical modeling of extreme values* (Vol. 208). Berlin: Springer.

Comrey, A. L., & Lee, H. B. (2013). *A first course in factor analysis*. UK: Psychology Press.

Cook, R., & Johnson, M. (1986). Generalized Burr-Pareto-logistic distributions with applications to a uranium exploration data set. *Technometrics, 28*(2), 123–131.

Cormen, T., Leiserson, E., Rivest, R., & Stein, C. (2009). *Introduction to algorithms*. Camebridge: The MIT Press.

Corsi, F., Mittnik, S., Pigorsch, C., & Pigorsch, U. (2008). The volatility of realized volatility. *Econometric Reviews, 27*(1–3), 46–78.

Cortez, P., Cerdeira, A., Almeida, F., Matos, T., & Reis, J. (2009). Modeling wine preferences by data mining from physicochemical properties. *Decision Support Systems, 47*, 547–553.

Czado, C., Jeske, S., & Hofmann, M. (2013). Selection strategies for regular vine copulae. *Journal de la Societe Francaise de Statistique, 154*(1), 172–191.

Czado, C., Kastenmeier, R., Brechmann, E. C., & Min, A. (2012). A mixed copula model for insurance claims and claim sizes. *Scandinavian Actuarial Journal, 2012*(4), 278–305.

Czado, C., Schepsmeier, U., & Min, A. (2012). Maximum likelihood estimation of mixed C-vines with application to exchange rates. *Statistical Modelling, 12*(3), 229–255.

de Haan, L., & Resnick, S. I. (1977). Limit theory for multivariate sample extremes. *Zeitschrift für Wahrscheinlichkeitstheorie und verwandte Gebiete, 40*(4), 317–337.

Demarta, S., & McNeil, A. J. (2005). The t copula and related copulas. *International Statistical Review, 73*(1), 111–129.

Denuit, M., & Lambert, P. (2005). Constraints on concordance measures in bivariate discrete data. *Journal of Multivariate Analysis, 93*(1), 40–57.

Diao, L., & Cook, R. J. (2014). Composite likelihood for joint analysis of multiple multistate processes via copulas. *Biostatistics, 15*(4), 690–705.

Diestel, R. (2006). *Graph theory* (3rd ed.). Berlin: Springer.

Diggle, P. (2002). *Analysis of longitudinal data*. Oxford: Oxford University Press.

Diggle, P. J., & Donnelly, J. B. (1989). A selected bibliography on the analysis of repeated measurements and related areas. *Australian Journal of Statistics, 31*(1), 183–193.

Dißmann, J., Brechmann, E. C., Czado, C., & Kurowicka, D. (2013). Selecting and estimating regular vine copulae and application to financial returns. *Computational Statistics and Data Analysis, 59*, 52–69.

Dißmann, J. F. (2010). Statistical inference for regular vines and application. Master's thesis, Technische Universität München.

Doléans-Dade, C., & Meyer, P.-A. (1970). Intégrales stochastiques par rapport aux martingales locales. *Séminaire de Probabilités IV Université de Strasbourg* (pp. 77–107). Berlin: Springer.

Durante, F., Girard, S., & Mazo, G. (2015). Copulas based on Marshall-Olkin machinery. *Marshall-Olkin Distributions: Advances in Theory and Applications* (pp. 15–31). Berlin: Springer.

Embrechts, P., Lindskog, F., & McNeal, A. (2003). *Modelling dependence with copulas and applications to risk management. In Handbook of Heavy Tailed Distributions in Finance*. Amsterdam: Elsevier/North-Holland.

Erhardt, T. M., & Czado, C. (2018). Standardized drought indices: a novel univariate and multivariate approach. *Journal of the Royal Statistical Society: Series C (Applied Statistics), 67*, 643–664.

Erhardt, T. M., Czado, C., & Schepsmeier, U. (2015a). R-vine models for spatial time series with an application to daily mean temperature. *Biometrics, 71*(2), 323–332.

Erhardt, T. M., Czado, C., & Schepsmeier, U. (2015b). Spatial composite likelihood inference using local C-vines. *Journal of Multivariate Analysis, 138*, 74–88.

Erhardt, V., & Czado, C. (2012). Modeling dependent yearly claim totals including zero claims in private health insurance. *Scandinavian Actuarial Journal, 2012*(2), 106–129.

Eschenburg, P. (2013). Properties of extreme-value copulas. Diploma thesis, Technische Universität München.

Fenske, N., Kneib, T., & Hothorn, T. (2012). Identifying risk factors for severe childhood malnutrition by boosting additive quantile regression. *Journal of the American Statistical Association, 106*, 494–510.

Fernández, C., & Steel, M. F. (1998). On Bayesian modeling of fat tails and skewness. *Journal of the American Statistical Association, 93*(441), 359–371.

Fink, H., Klimova, Y., Czado, C., & Stöber, J. (2017). Regime switching vine copula models for global equity and volatility indices. *Econometrics, 5*(1), 3.

Fischer, M., Kraus, D., Pfeuffer, M., & Czado, C. (2017). Stress testing German industry sectors: Results from a vine copula based quantile regression. *Risks, 5*(3), 38.

Fitzmaurice, G., Davidian, M., Verbeke, G., & Molenberghs, G. (2008). *Longitudinal data analysis*. USA: CRC Press.

Frahm, G., Junker, M., & Schmidt, R. (2005). Estimating the tail-dependence coefficient: properties and pitfalls. *Insurance: Mathematics and Economics, 37*(1), 80–100.

Francq, C., & Zakoian, J. (2011). *GARCH Models: Structure, Statistical Inference and Financial Applications*. New York: Wiley.

Friedman, J., Hastie, T., & Tibshirani, R. (2008). Sparse inverse covariance estimation with the graphical lasso. *Biostatistics, 9*(3), 432.

Galambos, J. (1975). Order statistics of samples from multivariate distributions. *Journal of the American Statistical Association, 70*(351a), 674–680.

Geenens, G., Charpentier, A., Paindaveine, D., et al. (2017). Probit transformation for nonparametric kernel estimation of the copula density. *Bernoulli, 23*(3), 1848–1873.

Genest, C., & Favre, A.-C. (2007). Everything you always wanted to know about copula modeling but were afraid to ask. *Journal of Hydrologic Engineering, 12*(4), 347–368.

Genest, C., Ghoudi, K., & Rivest, L. (1995). A semi-parametric estimation procedure of dependence parameters in multivariate families of distributions. *Biometrika, 82*, 543–552.

Genest, C., Masiello, E., & Tribouley, K. (2009). Estimating copula densities through wavelets. *Insurance: Mathematics and Economics, 44*(2), 170–181.

Genest, C., & Nešlehová, J. (2007). A primer on copulas for count data. *ASTIN Bulletin: The Journal of the IAA, 37*(2), 475–515.

Genest, C., & Nešlehová, J. (2013). Copula modeling for extremes. *Encyclopedia of Environmetrics*.

Genest, C., Quessy, J.-F., & Rémillard, B. (2006). Goodness-of-fit procedures for copula models based on the probability integral transformation. *Scandinavian Journal of Statistics, 33*(2), 337–366.

Genest, C., Rémillard, B., & Beaudoin, D. (2009). Goodness-of-fit tests for copulas: A review and a power study. *Insurance: Mathematics and Economics, 44*(2), 199–213.

Genton, M. (2004). *Skew-elliptical distributions and their applications: A journey beyond normality*. London: Taylor and Francis.

Gijbels, I., & Mielniczuk, J. (1990). Estimating the density of a copula function. *Communications in Statistics - Theory and Methods, 19*(2), 445–464.

Gijbels, I., Omelka, M., Veraverbeke, N., et al. (2015). Partial and average copulas and association measures. *Electronic Journal of Statistics, 9*(2), 2420–2474.

Gilks, W. R., Richardson, S., & Spiegelhalter, D. (1995). *Markov chain Monte Carlo in practice*. USA: CRC Press.

Gräler, B. (2014). Modelling skewed spatial random fields through the spatial vine copula. *Spatial Statistics, 10*, 87–102.

Gräler, B., & Pebesma, E. (2011). The pair-copula construction for spatial data: a new approach to model spatial dependency. *Procedia Environmental Sciences, 7*, 206–211.

Green, P. J. (1995). Reversible jump Markov chain Monte Carlo computation and Bayesian model determination. *Biometrika, 82*(4), 711–732.

Gruber, L., & Czado, C. (2015). Sequential Bayesian model selection of regular vine copulas. *Bayesian Analysis, 10*(4), 937–963.

Gruber, L., & Czado, C. (2018). Bayesian model selection of regular vine copulas. *Bayesian Analysis, 13*(4), 1111–1135.

Gudendorf, G., & Segers, J. (2010). Extreme-value copulas. *Copula Theory and Its Applications* (pp. 127–145). Berlin: Springer.

Gurgul, H., & Machno, A. (2016). Modeling dependence structure among European markets and among Asian-Pacific markets: a regime switching regular vine copula approach. *Central European Journal of Operations Research, 24*(3), 763–786.

Haff, I. H., et al. (2013). Parameter estimation for pair-copula constructions. *Bernoulli, 19*(2), 462–491.

Haff, I. H., Aas, K., Frigessi, A., & Lacal, V. (2016). Structure learning in Bayesian networks using regular vines. *Computational Statistics and Data Analysis, 101*, 186–208.

Hafner, C. M., & Manner, H. (2012). Dynamic stochastic copula models: Estimation, inference and applications. *Journal of Applied Econometrics, 27*(2), 269–295.

Hahsler, M., & Hornik, K. (2017). TSP: Traveling sales person problem (TSP) R package version 1. 1–5. https://CRAN.R-project.org/package=TSP.

Hajek, J., & Sidak, Z. (1967). *Theory of rank statistics*. Academia Praha.

Han, D., Tan, K. S., & Weng, C. (2017). Vine copula models with GLM and sparsity. *Communications in Statistics - Theory and Methods, 46*(13), 6358–6381.

Hanea, A., Napoles, O. M., & Ababei, D. (2015). Non-parametric Bayesian networks: Improving theory and reviewing applications. *Reliability Engineering and System Safety, 144*, 265–284.

Hanea, A. M., Kurowicka, D., & Cooke, R. M. (2006). Hybrid method for quantifying and analyzing Bayesian belief nets. *Quality and Reliability Engineering International, 22*(6), 709–729.

Hobæk Haff, I., Frigessi, A., & Maraun, D. (2015). How well do regional climate models simulate the spatial dependence of precipitation? An application of pair-copula constructions. *Journal of Geophysical Research: Atmospheres, 120*(7), 2624–2646.

Hoeffding, W. (1948). A class of statistics with asymptotically normal distribution. *The Annals of Mathematical Statistics, 19*(3), 293–325.

Hofert, M., Kojadinovic, I., Maechler, M., & Yan, J. (2017). copula: Multivariate Dependence with Copulas. R package version 0.999–17.

Hoffman, M. D., & Gelman, A. (2014). The no-u-turn sampler: Adaptively setting path lengths in Hamiltonian Monte Carlo. *Journal of Machine Learning Research, 15*(1), 1593–1623.

Höhndorf, L., Czado, C., Bian, H., Kneer, J., & Holzapfel, F. (2017). Statistical modeling of dependence structures of operational flight data measurements not fulfilling the iid condition. In *AIAA Atmospheric Flight Mechanics Conference* (p. 3395).

Hollander, M., Wolfe, D. A., & Chicken, E. (2014). *Nonparametric statistical methods*. New York: Wiley.

Hu, Z., & Mahadevan, S. (2017). Time-dependent reliability analysis using a Vine-ARMA load model. *ASCE-ASME Journal of Risk and Uncertainty in Engineering Systems, Part B: Mechanical Engineering, 3*(1), 011007.

Huang, W., & Prokhorov, A. (2014). A goodness-of-fit test for copulas. *Econometric Reviews, 33*(7), 751–771.

Hürlimann, W. (2003). Hutchinson-Lai's conjecture for bivariate extreme value copulas. *Statistics and Probability letters, 61*(2), 191–198.

Hüsler, J., & Reiss, R.-D. (1989). Maxima of normal random vectors: between independence and complete dependence. *Statistics and Probability Letters, 7*(4), 283–286.

Ivanov, E., Min, A., & Ramsauer, F. (2017). Copula-based factor models for multivariate asset returns. *Econometrics, 5*(2), 20.

Jacod, J. (1994). *Limit of random measures associated with the increments of a brownian semi-martingale, 120*, 155–162.

Jäger, W., & Nápoles, O. M. (2017). A vine-copula model for time series of significant wave heights and mean zero-crossing periods in the North Sea. *ASCE-ASME Journal of Risk and Uncertainty in Engineering Systems, Part A: Civil Engineering, 3*(4), 04017014.

Jiang, C., Zhang, W., Han, X., Ni, B., & Song, L. (2015). A vine-copula-based reliability analysis method for structures with multidimensional correlation. *Journal of Mechanical Design, 137*(6), 061405.

Joe, H. (1996). Families of m-variate distributions with given margins and m(m-1)/2 bivariate dependence parameters. In L. Rüschendorf, B. Schweizer, & M. D. Taylor (Eds.), *Distributions with Fixed Marginals and Related Topics.*

Joe, H. (1997). *Multivariate models and multivariate dependence concepts.* USA: CRC Press.

Joe, H. (2005). Asymptotic efficiency of the two stage estimation method for copula-based models. *Journal of Multivariate Analysis, 94*, 401–419.

Joe, H. (2006). Generating random correlation matrices based on partial correlations. *Journal of Multivariate Analysis, 97*(10), 2177–2189.

Joe, H. (2014). *Dependence modeling with copulas.* USA: CRC Press.

Joe, H., Li, H., & Nikoloulopoulos, A. K. (2010). Tail dependence functions and vine copulas. *Journal of Multivariate Analysis, 101*(1), 252–270.

Joe, H. & Xu, J. (1996). The estimation method of inference functions for margins for multivariate models. Technical Report 166, Department of Statistics, University of British Columbia.

Johnson, N. L., Kotz, S., & Balakrishnan, N. (1995). *Continuous univariate distributions* (Vol. 2). New York: Wiley.

Kaplan, D. (2009). *Structural equation modeling: foundations and extensions* (2nd ed.). Thousand Oaks, California: SAGE Publications.

Kauermann, G., & Schellhase, C. (2014). Flexible pair-copula estimation in D-vines using bivariate penalized splines. *Statistics and Computing, 24*(6), 1081–1100.

Kauermann, G., Schellhase, C., & Ruppert, D. (2013). Flexible copula density estimation with penalized hierarchical B-splines. *Scandinavian Journal of Statistics, 40*, 685–705.

Killiches, M., & Czado, C. (2015). Block-maxima of vines. In D. Dey & J. Yan (Eds.), *Extreme Value Modelling and Risk Analysis: Methods and Applications* (pp. 109–130). FL: Chapman & Hall/CRC Press.

Killiches, M., & Czado, C. (2018). A D-vine copula based model for repeated measurements extending linear mixed models with homogeneous correlation structure. *Biometrics, 74*(3), 997–1005.

Killiches, M., Kraus, D., & Czado, C. (2017). Examination and visualisation of the simplifying assumption for vine copulas in three dimensions. *Australian and New Zealand Journal of Statistics, 59*(1), 95–117.

Killiches, M., Kraus, D., & Czado, C. (2018). Model distances for vine copulas in high dimensions. *Statistics and Computing, 28*(2), 323–341.

Kim, G., Silvapulle, M. J., & Silvapulle, P. (2007). Comparison of semiparametric and parametric methods for estimating copulas. *Computational Statistics and Data Analysis, 51*(6), 2836–2850.

Klugman, S. A., Panjer, H. H., & Willmot, G. E. (2012). *Loss models: from data to decisions* (Vol. 715). New York: Wiley.

Knight, W. R. (1966). A computer method for calculating Kendall's tau with ungrouped data. *Journal of the American Statistical Association, 61*(314), 436–439.

Koenker, R., & Bassett, G. (1978). Regression quantiles. *Econometrica: Journal of the Econometric Society, 46*, 33–50.

Koller, D., & Friedman, N. (2009). *Probabilistic graphical models: principles and techniques* (1st ed.). Cambridge: MIT Press.

Koopman, S. J., Lit, R., & Lucas, A. (2015). Intraday stock price dependence using dynamic discrete copula distributions. Tinbergen Institute Discussion Paper, No. 15-037/III/DSF90, Tinbergen Institute, Amsterdam and Rotterdam.

Kosgodagan, A. (2017). *High-dimensional dependence modelling using Bayesian networks for the degradation of civil infrastructures and other applications.* Ph. D. thesis, Ecole nationale supérieure Mines-Télécom Atlantique.

Kotz, S., & Nadarajah, S. (2004). *Multivariate t distributions and their applications.* Cambridge: Cambridge University Press.

Krämer, N., Brechmann, E. C., Silvestrini, D., & Czado, C. (2013). Total loss estimation using copula-based regression models. *Insurance: Mathematics and Economics, 53*(3), 829–839.

Kraus, D., & Czado, C. (2017a). D-vine copula based quantile regression. *Computational Statistics and Data Analysis, 110C*, 1–18.

Kraus, D., & Czado, C. (2017b). Growing simplified vine copula trees: improving Dißmann's algorithm. arXiv:1703.05203.

Kreuzer, A., & Czado, C. (2018). Bayesian inference for a single factor copula stochastic volatility model using Hamiltonian Monte Carlo. arXiv:1808.08624.

Krupskii, P., Huser, R., & Genton, M. G. (2018). Factor copula models for replicated spatial data. *Journal of the American Statistical Association, 113*(521), 467–479.

Krupskii, P., & Joe, H. (2013). Factor copula models for multivariate data. *Journal of Multivariate Analysis, 120*, 85–101.

Krupskii, P., & Joe, H. (2015a). Structured factor copula models: Theory, inference and computation. *Journal of Multivariate Analysis, 138*, 53–73.

Krupskii, P., & Joe, H. (2015b). Tail-weighted measures of dependence. *Journal of Applied Statistics, 42*(3), 614–629.

Kullback, S., & Leibler, R. A. (1951). On information and sufficiency. *The Annals of Mathematical Statistics, 22*(1), 79–86.

Kurowicka, D. (2009). Some results for different strategies to choose optimal vine truncation based on wind speed data. In *Conference presentation, 3rd Vine Copula Workshop, Oslo*.

Kurowicka, D. (2011). Optimal truncation of vines. In D. Kurowicka & H. Joe (Eds.), *Dependence Modeling-handbook on Vine Copulae*. Singapore: World Scientific Publishing Co.

Kurowicka, D., & Cooke, R. (2006). *Uncertainty analysis with high dimensional dependence modelling.* Chichester: Wiley.

Kurowicka, D., & Cooke, R. M. (2005). Sampling algorithms for generating joint uniform distributions using the vine - copula method. In *3rd IASC World Conference on Computational Statistics and Data Analysis, Limassol, Cyprus*.

Kurowicka, D., & Joe, H. (2011). *Dependence modeling - handbook on vine copulae.* Singapore: World Scientific Publishing Co.

Kurz, M. (2015). A MATLAB toolbox for vine copulas based on C++. MATLAB toolbox.

Kurz, M. S. (2017). Pacotest: Testing for Partial Copulas and the Simplifying Assumption in Vine Copulas. R package version 0.2.2.

Kurz, M. S., & Spanhel, F. (2017). Testing the simplifying assumption in high-dimensional vine copulas. arXiv:1706.02338.

Laevens, H., Deluyker, H., Schukken, Y. H., De Meulemeester, L., Vandermeersch, R., De Meulenaere, E., et al. (1997). Influence of parity and stage of lactation on the somatic cell count in bacteriologically negative dairy cows. *Journal of Dairy Science, 80*, 3219–3226.

Lambert, P., & Vandenhende, F. (2002). A copula-based model for multivariate non-normal longitudinal data: analysis of a dose titration safety study on a new antidepressant. *Statistics in Medicine*, *21*(21), 3197–3217.

Lauritzen, S. L. (1996). *Graphical Models* (1st ed.). England: University Press.

Lee, D., & Joe, H. (2018). Multivariate extreme value copulas with factor and tree dependence structures. *Extremes*, *21*(1), 147–176.

Lehmann, E. L., & Casella, G. (2006). *Theory of point estimation*. Berlin: Springer Science & Business Media.

Li, D. X. (2000). On default correlation. *The Journal of Fixed Income*, *9*(4), 43–54.

Li, Q., Lin, J., & Racine, J. S. (2013). Optimal bandwidth selection for nonparametric conditional distribution and quantile functions. *Journal of Business and Economic Statistics*, *31*(1), 57–65.

Mai, J.-F., & Scherer, M. (2010). The Pickands representation of survival Marshall-Olkin copulas. *Statistics and Probability Letters*, *80*(5), 357–360.

Manner, H. (2007). Estimation and model selection of copulas with an application to exchange rates. METEOR research memorandum 07/056, Maastricht University.

Manner, H., Fard, F. A., Pourkhanali, A., & Tafakori, L. (2019). Forecasting the joint distribution of Australian electricity prices using dynamic vine copulae. *Energy Economics*, *78*, 143–164.

Manner, H., & Reznikova, O. (2012). A survey on time-varying copulas: specification, simulations, and application. *Econometric Reviews*, *31*(6), 654–687.

Marshall, A. W., & Olkin, I. (1967). A multivariate exponential distribution. *Journal of the American Statistical Association*, *62*(317), 30–44.

Massey, F. J, Jr. (1951). The Kolmogorov-Smirnov test for goodness of fit. *Journal of the American Statistical Association*, *46*(253), 68–78.

McNeil, A. J., Frey, R., & Embrechts, P. (2015). *Quantitative risk management: Concepts, techniques and tools*. Princeton: Princeton University Press.

McNeil, A. J., & Nešlehová, J. (2009). Multivariate Archimedean copulas, d-monotone functions and ℓ_1-norm symmetric distributions. *Annals of Statistics*, *37*(5B), 3059–3097.

Meester, S. G., & MacKay, J. (1994). A parametric model for cluster correlated categorical data. *Biometrics*, *50*(4), 954–963.

Mendes, B Vd, & M, Semeraro, M. M., & Leal, R. P., (2010). Pair-copulas modeling in finance. *Financial Markets and Portfolio Management*, *24*(2), 193–213.

Metropolis, N., Rosenbluth, A. W., Rosenbluth, M. N., Teller, A. H., & Teller, E. (1953). Equation of state calculations by fast computing machines. *The Journal of Chemical Physics*, *21*(6), 1087–1092.

Min, A., & Czado, C. (2010). Bayesian inference for multivariate copulas using pair-copula constructions. *Journal of Financial Econometrics*, *8*(4), 511–546.

Min, A., & Czado, C. (2011). Bayesian model selection for D-vine pair-copula constructions. *Canadian Journal of Statistics*, *39*(2), 239–258.

Möller, A., Spazzini, L., Kraus, D., Nagler, T., & Czado, C. (2018). Vine copula based postprocessing of ensemble forecasts for temperature. arXiv:1811.02255.

Morales-Nápoles, O. (2011). Counting vines. In D. Kurowicka & H. Joe (Eds.), *Dependence modeling: vine copula handbook*. Singapore: World Scientific Publishing Co.

Müller, D., & Czado, C. (2017). Dependence modeling in ultra high dimensions with vine copulas and the graphical lasso. arXiv:1709.05119.

Müller, D., & Czado, C. (2018a). Representing sparse Gaussian DAGs as sparse R-vines allowing for non-Gaussian dependence. *Journal of Computational and Graphical Statistics*, *27*(2), 334–344.

Müller, D., & Czado, C. (2018b). Selection of sparse vine copulas in high dimensions with the lasso. *Statistics and Computing first online*, 1–19.

Nagler, T. (2014). Kernel methods for vine copula estimation. Master thesis, Technische Universität München.

Nagler, T. (2016). kdecopula: An R package for the kernel estimation of copula densities. arXiv:1603.04229.

Nagler, T. (2017a). kdecopula: Kernel Smoothing for Bivariate Copula Densities. *R package version*, (9),

Nagler, T. (2017b). kdevine: Multivariate Kernel Density Estimation with Vine Copulas. *R package version*, (4), 1.

Nagler, T. (2018). Asymptotic analysis of the jittering kernel density estimator. *Mathematical Methods of Statistics*, *27*(1), 32–46.

Nagler, T., Bumann, C., & Czado, C. (2018). Model selection in sparse high-dimensional vine copula models with application to portfolio risk. arXiv:1801.09739.

Nagler, T., & Czado, C. (2016). Evading the curse of dimensionality in nonparametric density estimation with simplified vine copulas. *Journal of Multivariate Analysis*, *151*, 69–89.

Nagler, T., & Kraus, D. (2017). vinereg: D-Vine Quantile Regression. *Version*, (1), 3.

Nagler, T., Schellhase, C., & Czado, C. (2017). Nonparametric estimation of simplified vine copula models: comparison of methods. *Dependence Modeling*, *5*, 99–120.

Nagler, T., & Vatter, T. (2017). rvinecopulib: High Performance Algorithms for Vine Copula Modeling. R package version 0.1.0.1.1.

Nai Ruscone, M., & Osmetti, S. A. (2017). Modelling the dependence in multivariate longitudinal data by pair copula decomposition. In Ferraro, M. B., Giordani, P., Vantaggi, B., Gagolewski, M., Angeles Gil, M., Grzegorzewski, P., Hryniewicz, O. (Eds.), *Soft Methods for Data Science* (pp. 373–380). Springer International Publishing Switzerland.

Nelsen, R. (2006). *An introduction to copulas*. New York: Springer.

Nikoloulopoulos, A. K. (2013). On the estimation of normal copula discrete regression models using the continuous extension and simulated likelihood. *Journal of Statistical Planning and Inference*, *143*(11), 1923–1937.

Nikoloulopoulos, A. K. (2017). A vine copula mixed effect model for trivariate meta-analysis of diagnostic test accuracy studies accounting for disease prevalence. *Statistical Methods in Medical Research*, *26*(5), 2270–2286.

Nikoloulopoulos, A. K., & Joe, H. (2015). Factor copula models for item response data. *Psychometrika*, *80*(1), 126–150.

Nikoloulopoulos, A. K., Joe, H., & Li, H. (2012). Vine copulas with asymmetric tail dependence and applications to financial return data. *Computational Statistics and Data Analysis*, *56*(11), 3659–3673.

Oh, D. H., & Patton, A. J. (2017). Modeling dependence in high dimensions with factor copulas. *Journal of Business and Economic Statistics*, *35*(1), 139–154.

Okhrin, O., Okhrin, Y., & Schmid, W. (2013). On the structure and estimation of hierarchical Archimedean copulas. *Journal of Econometrics*, *173*(2), 189–204.

Panagiotelis, A., Czado, C., & Joe, H. (2012). Pair copula constructions for multivariate discrete data. *Journal of the American Statistical Association*, *107*(499), 1063–1072.

Pearson, K. (1904). *Mathematical contributions to the theory of evolution*. Dulau and co.

Pereira, G., Veiga, Á., Erhardt, T., & Czado, C. (2016). Spatial R-vine copula for streamflow scenario simulation. In *Power Systems Computation Conference (PSCC), 2016* (pp. 1–7). IEEE.

Pereira, G. A., Veiga, Á., Erhardt, T., & Czado, C. (2017). A periodic spatial vine copula model for multi-site streamflow simulation. *Electric Power Systems Research*, *152*, 9–17.

Pickands, J. (1981). Multivariate extreme value distributions. *Proceedings 43rd Session International Statistical Institute* (Vol. 2, pp. 859–878).

Pivotal Inc. (2017). PivotalR: A Fast, Easy-to-Use Tool for Manipulating Tables in Databases and a Wrapper of MADlib. R package version 0.1.18.3.

Puccetti, G., & Scherer, M. (2018). Copulas, credit portfolios, and the broken heart syndrome. *Dependence Modeling*, *6*(1). published online.

Core Team, R. (2017). *R: A Language and Environment for Statistical Computing*. Austria: R Foundation for Statistical Computing.

Rémillard, B., Nasri, B., & Bouezmarni, T. (2017). On copula-based conditional quantile estimators. *Statistics and Probability Letters, 128*, 14–20.

Ripley, B. D. (2008). *Pattern recognition and neural networks*. Cambridge: Cambridge University Press.

Robert, C. P. (2004). *Monte Carlo methods*. New York: Wiley Online Library.

Rosenblatt, M. (1952). Remarks on multivariate transformation. *Annals of Mathematical Statistics, 23*, 1052–1057.

Ruppert, D. (2004). *Statistics and finance: an introduction*. Berlin: Springer Science & Business Media.

Rüschendorf, L. (1981). Stochastically ordered distributions and monotonicity of the oc-function of sequential probability ratio tests. *Statistics: A Journal of Theoretical and Applied Statistics, 12*(3), 327–338.

Sachs, K., Omar, P., Pe'er, D., Lauffenburger, D. A., & Nolan, G. P. (2005). Causal protein signaling networks derived from multiparameter single-cell data. *Science, 308*, 523–529.

Salmon, F. (2012). The formula that killed wall street. *Significance, 9*(1), 16–20.

Sancetta, A., & Satchell, S. (2004). The Bernstein copula and its applications to modeling and approximations of multivariate distributions. *Econometric Theory, 20*(3), 535–562.

Savu, C., & Trede, M. (2010). Hierarchies of Archimedean copulas. *Quantitative Finance, 10*(3), 295–304.

Schallhorn, N., Kraus, D., Nagler, T., & Czado, C. (2017). D-vine quantile regression with discrete variables. arXiv:1705.08310.

Schamberger, B., Gruber, L. F., & Czado, C. (2017). Bayesian inference for latent factor copulas and application to financial risk forecasting. *Econometrics, 5*(2), 21.

Schellhase, C. (2017a). pencopulaCond: Estimating Non-Simplified Vine Copulas Using Penalized Splines. *R package version, 2*.

Schellhase, C. (2017b). penRvine: Flexible R-Vines Estimation Using Bivariate Penalized Splines. *R package version, 2*.

Schellhase, C., & Spanhel, F. (2018). Estimating non-simplified vine copulas using penalized splines. *Statistics and Computing, 28*(2), 387–409.

Schepsmeier, U. (2014). *Estimating standard errors and efficient goodness-of-fit tests for regular vine copula models*. Ph. D. thesis, Technische Universität München.

Schepsmeier, U. (2015). Efficient information based goodness-of-fit tests for vine copula models with fixed margins: A comprehensive review. *Journal of Multivariate Analysis, 138*, 34–52.

Schepsmeier, U. (2016). A goodness-of-fit test for regular vine copula models. *Econometric Reviews*, 1–22.

Schepsmeier, U., & Czado, C. (2016). Dependence modelling with regular vine copula models: a case-study for car crash simulation data. *Journal of the Royal Statistical Society: Series C (Applied Statistics), 65*(3), 415–429.

Schepsmeier, U., & Stöber, J. (2014). Derivatives and fisher information of bivariate copulas. *Statistical Papers, 55*(2), 525–542.

Schepsmeier, U., Stöber, J., Brechmann, E. C., Gräler, B., Nagler, T., & Erhardt, T. (2018). VineCopula: statistical inference of vine copulas. *Version, 2*(1), 6.

Schmidl, D., Czado, C., Hug, S., Theis, F. J., et al. (2013). A vine-copula based adaptive MCMC sampler for efficient inference of dynamical systems. *Bayesian Analysis, 8*(1), 1–22.

Schwarz, G. (1978). Estimating the dimension of a model. *The Annals of Statistics, 6*(2), 461–464.

Scott, D. W. (2015). The curse of dimensionality and dimension reduction. *Multivariate Density Estimation* (pp. 217–240). New York: Wiley.

Scutari, M. (2010). Learning Bayesian Networks with the bnlearn R package. *Journal of Statistical Software, 35*(3), 1–22.

Shen, C., & Weissfeld, L. (2006). A copula model for repeated measurements with non-ignorable non-monotone missing outcome. *Statistics in Medicine, 25*(14), 2427–2440.

Shi, P., Feng, X., Boucher, J.-P., et al. (2016). Multilevel modeling of insurance claims using copulas. *The Annals of Applied Statistics, 10*(2), 834–863.

Shi, P., & Yang, L. (2018). Pair copula constructions for insurance experience rating. *Journal of the American Statistical Association, 113*(521), 122–133.

Sklar, A. (1959). Fonctions de répartition à n dimensions et leurs marges. *Publications de l'Institut de Statistique de L'Université de Paris, 8*, 229–231.

Smith, M., Min, A., Almeida, C., & Czado, C. (2010). Modeling longitudinal data using a pair-copula decomposition of serial dependence. *Journal of the American Statistical Association, 105*(492), 1467–1479.

Smith, M. S. (2015). Copula modelling of dependence in multivariate time series. *International Journal of Forecasting, 31*(3), 815–833.

Spanhel, F., & Kurz, M. S. (2015). Simplified vine copula models: Approximations based on the simplifying assumption. arXiv:1510.06971.

Spirtes, P., Glymour, C., Scheines, R., Kauffman, S., Aimale, V., & Wimberly, F. (2000). Constructing Bayesian network models of gene expression networks from microarray data.

Spissu, E., Pinjari, A. R., Pendyala, R. M., & Bhat, C. R. (2009). A copula-based joint multinomial discrete-continuous model of vehicle type choice and miles of travel. *Transportation, 36*(4), 403–422.

Stöber, J. (2013). *Regular Vine Copulas with the simplifying assumption, time-variation, and mixed discrete and continuous margins*. Ph.D. thesis, Technische Universität München.

Stöber, J., & Czado, C. (2014). Regime switches in the dependence structure of multidimensional financial data. *Computational Statistics and Data Analysis, 76*, 672–686.

Stöber, J., & Czado, C. (2017). Pair copula constructions. In *Simulating Copulas: Stochastic Models, Sampling Algorithms, and Applications*, (2nd ed., pp. 185–230). Singapore: World Scientific Publishing.

Stöber, J., Hong, H. G., Czado, C., & Ghosh, P. (2015). Comorbidity of chronic diseases in the elderly: Patterns identified by a copula design for mixed responses. *Computational Statistics and Data Analysis, 88*, 28–39.

Stöber, J., Joe, H., & Czado, C. (2013). Simplified pair copula constructions-limitations and extensions. *Journal of Multivariate Analysis, 119*, 101–118.

Stöber, J., & Schepsmeier, U. (2013). Estimating standard errors in regular vine copula models. *Computational Statistics, 28*(6), 2679–2707.

Sun, J., Frees, E. W., & Rosenberg, M. A. (2008). Heavy-tailed longitudinal data modeling using copulas. *Insurance: Mathematics and Economics, 42*(2), 817–830.

Takahashi, K. (1965). Note on the multivariate Burr's distribution. *Annals of the Institute of Statistical Mathematics, 17*, 257–260.

Tawn, J. A. (1988). Bivariate extreme value theory: models and estimation. *Biometrika, 75*(3), 397–415.

Tibshirani, R. (1996). Regression shrinkage and selection via the lasso. *Journal of the Royal Statistical Society. Series B (Methodological), 58*(1), 267–288.

Timmer, R., Broussard, J. P., & Booth, G. G. (2018). The efficacy of life insurance company general account equity asset allocations: a safety-first perspective using vine copulas. *Annals of Actuarial Science, 12*(2), 372–390.

Van der Waerden, B. (1953). Ein neuer Test für das Problem der zwei Stichproben. *Mathematische Annalen, 126*(1), 93–107.

Vatter, T., & Chavez-Demoulin, V. (2015). Generalized additive models for conditional dependence structures. *Journal of Multivariate Analysis, 141*, 147–167.

Vatter, T., & Nagler, T. (2017). GamCopula: Generalized Additive Models for Bivariate Conditional Dependence Structures and Vine Copulas. R package version 0.0-4.

Vatter, T., & Nagler, T. (2018). Generalized additive models for pair-copula constructions. *Journal of Computational and Graphical Statistics in press*, 1–13. online available.

Verbeke, G., & Molenberghs, G. (2009). *Linear mixed models for longitudinal data*. Berlin: Springer Science & Business Media.

Vuong, Q. H. (1989). Likelihood ratio tests for model selection and non-nested hypotheses. *Econometrica*, *57*, 307–333.

White, H. (1982). Maximum likelihood estimation of misspecified models. *Econometrica: Journal of the Econometric Society*, *50*(1), 1–25.

Wuertz, D., Setz, T., Chalabi, Y., Boudt, C., Chausse, P., & Miklovac, M. (2017). fGarch: Rmetrics - Autoregressive Conditional Heteroskedastic Modelling. *R Package Version*, *3042*, 83.

Yule, U., & Kendall, M. (1950). *An introduction to the theory of statistics* (14th ed.). London: Charles Griffin & Co., Ltd.

Zhou, Q. M., Song, P. X.-K., & Thompson, M. E. (2012). Information ratio test for model misspecification in quasi-likelihood inference. *Journal of the American Statistical Association*, *107*(497), 205–213.

Zimmer, D. M., & Trivedi, P. K. (2006). Using trivariate copulas to model sample selection and treatment effects: application to family health care demand. *Journal of Business and Economic Statistics*, *24*(1), 63–76.

Index

© Springer Nature Switzerland AG 2019
C. Czado, *Analyzing Dependent Data with Vine Copulas*, Lecture Notes
in Statistics 222, https://doi.org/10.1007/978-3-030-13785-4

Printed in the United States
By Bookmasters